FOCUS ON
GEODATABASES

in ArcGIS® Pro

David W. Allen

Esri Press
REDLANDS | CALIFORNIA

Esri Press, 380 New York Street, Redlands, California 92373-8100
Copyright 2019 Esri
All rights reserved.
23 22 21 20 19   1 2 3 4 5 6 7 8 9 10
Printed in the United States of America

Library of Congress Cataloging-in-Publication Data

Names: Allen, David W., 1961- author.
Title: Focus on geodatabases in ArcGIS Pro / David W. Allen.
Description: Redlands, California : Esri Press, [2019]
Identifiers: LCCN 2019001552 (print) | LCCN 2019014319 (ebook) | ISBN 9781589484467 (electronic) | ISBN 9781589484450 (pbk. : alk. paper)
Subjects: LCSH: ArcGIS. | Geodatabases.
Classification: LCC G70.212 (ebook) | LCC G70.212 .A432 2019 (print) | DDC 910.285/53--dc23
LC record available at https://urldefense.proofpoint.com/v2/url?u=https-3A__lccn.loc.gov_2019001552&d=DwIFAg&c=n6-cguzQvX_tUIrZOS_4Og&r=qNU49__SCQN30XC-f38qj8bYYMTIH4VCOt-Jb8fvjUA&m=qtrrm-_9q_bdtX5eqLjFW5R16kCSjLoGvlCLUmQJLwE&s=L04HzaZUFMmLqGNqKNxL6TeShnG4d_7z1ZXl6smHq1Q&e=

The information contained in this document is the exclusive property of Esri unless otherwise noted. This work is protected under United States copyright law and the copyright laws of the given countries of origin and applicable international laws, treaties, and/or conventions. No part of this work may be reproduced or transmitted in any form or by any means, electronic or mechanical, including photocopying or recording, or by any information storage or retrieval system, except as expressly permitted in writing by Esri. All requests should be sent to Attention: Contracts and Legal Services Manager, Esri, 380 New York Street, Redlands, California 92373-8100, USA.

The information contained in this document is subject to change without notice.

US Government Restricted/Limited Rights: Any software, documentation, and/or data delivered hereunder is subject to the terms of the License Agreement. The commercial license rights in the License Agreement strictly govern Licensee's use, reproduction, or disclosure of the software, data, and documentation. In no event shall the US Government acquire greater than RESTRICTED/LIMITED RIGHTS. At a minimum, use, duplication, or disclosure by the US Government is subject to restrictions as set forth in FAR §52.227-14 Alternates I, II, and III (DEC 2007); FAR §52.227-19(b) (DEC 2007) and/or FAR §12.211/12.212 (Commercial Technical Data/Computer Software); and DFARS §252.227-7015 (DEC 2011) (Technical Data – Commercial Items) and/or DFARS §227.7202 (Commercial Computer Software and Commercial Computer Software Documentation), as applicable. Contractor/Manufacturer is Esri, 380 New York Street, Redlands, CA 92373-8100, USA.

@esri.com, 3D Analyst, ACORN, Address Coder, ADF, AML, ArcAtlas, ArcCAD, ArcCatalog, ArcCOGO, ArcData, ArcDoc, ArcEdit, ArcEditor, ArcEurope, ArcExplorer, ArcExpress, ArcGIS, arcgis.com, ArcGlobe, ArcGrid, ArcIMS, ARC/INFO, ArcInfo, ArcInfo Librarian, ArcLessons, ArcLocation, ArcLogistics, ArcMap, ArcNetwork, *ArcNews*, ArcObjects, ArcOpen, ArcPad, ArcPlot, ArcPress, ArcPy, ArcReader, ArcScan, ArcScene, ArcSchool, ArcScripts, ArcSDE, ArcSdl, ArcSketch, ArcStorm, ArcSurvey, ArcTIN, ArcToolbox, ArcTools, ArcUSA, *ArcUser*, ArcView, ArcVoyager, *ArcWatch*, ArcWeb, ArcWorld, ArcXML, Atlas GIS, AtlasWare, Avenue, BAO, Business Analyst, Business Analyst Online, BusinessMAP, CityEngine, CommunityInfo, Database Integrator, DBI Kit, EDN, Esri, esri.com, Esri—Team GIS, Esri—*The GIS Company*, Esri—The GIS People, Esri—The GIS Software Leader, FormEdit, GeoCollector, Geographic Design System, Geography Matters, Geography Network, geographynetwork.com, Geoloqi, Geotrigger, GIS by Esri, gis.com, GISData Server, GIS Day, gisday.com, GIS for Everyone, JTX, MapIt, Maplex, MapObjects, MapStudio, ModelBuilder, MOLE, MPS—Atlas, PLTS, Rent-a-Tech, SDE, SML, Sourcebook•America, SpatiaLABS, Spatial Database Engine, StreetMap, Tapestry, the ARC/INFO logo, the ArcGIS Explorer logo, the ArcGIS logo, the ArcPad logo, the Esri globe logo, the Esri Press logo, The Geographic Advantage, The Geographic Approach, the GIS Day logo, the MapIt logo, The World's Leading Desktop GIS, *Water Writes*, and Your Personal Geographic Information System are trademarks, service marks, or registered marks of Esri in the United States, the European Community, or certain other jurisdictions. CityEngine is a registered trademark of Procedural AG and is distributed under license by Esri. Other companies and products or services mentioned herein may be trademarks, service marks, or registered marks of their respective mark owners.

Ask for Esri Press titles at your local bookstore or order by calling 1-800-447-9778. You can also shop online at www.esri.com/esripress. Outside the United States, contact your local Esri distributor or shop online at eurospanbookstore.com/esri.

Esri Press titles are distributed to the trade by the following:

In North America:
Ingram Publisher Services
Toll-free telephone: 800-648-3104
Toll-free fax: 800-838-1149
E-mail: customerservice@ingrampublisherservices.com

In the United Kingdom, Europe, the Middle East and Africa, Asia, and Australia:
Eurospan Group   Telephone 44(0) 1767 604972
3 Henrietta Street   Fax: 44(0) 1767 6016-40
London WC2E 8LU   E-mail: eurospan@turpin-distribution.com
United Kingdom

Contents

Introduction vii

Chapter 1   **Designing the geodatabase schema** 1
     Tutorial 1-1: Creating a geodatabase—building a logical model 1
         Exercise 1-1 19
     Tutorial 1-2: Creating a geodatabase—expanding the logical model 21
         Exercise 1-2 29

Chapter 2   **Creating a geodatabase** 33
     Tutorial 2-1: Building a geodatabase 33
         Exercise 2-1 61
     Tutorial 2-2: Adding complex geodatabase components 64
         Exercise 2-2 74

Chapter 3   **Populating and sharing a geodatabase** 79
     Tutorial 3-1: Loading data into a geodatabase 79
         Exercise 3-1 91
     Tutorial 3-2: Populating geodatabase subtypes 93
         Exercise 3-2 103

Chapter 4   **Extending data formats** 107
     Tutorial 4-1: Putting your data online 107
         Exercise 4-1 120
     Tutorial 4-2: Creating 3D scenes 122
         Exercise 4-2 131

Chapter 5   **Working with features** 133
     Tutorial 5-1: Creating new features 133
         Exercise 5-1 150
     Tutorial 5-2: Using context menu creation tools 153
         Exercise 5-2 166

Chapter 6   **Advanced editing** 169
     Tutorial 6-1: Exploring more creation tools 169
         Exercise 6-1 181
     Tutorial 6-2: Building feature templates 183
         Exercise 6-2 212

Chapter 7   **Working with topology** 215
     Tutorial 7-1: Working with map topology 215
         Exercise 7-1 223
     Tutorial 7-2: Working with geodatabase topology 224
         Exercise 7-2 244

Index 247

Introduction

The ArcGIS® platform uses the geodatabase as the spatial data format of choice for desktop, enterprise, web, and mobile applications. Within the geodatabase, you can store points, lines, polygons, tables, and other features for practically any application. As spatial data becomes more prolific across many platforms, the issue of maintaining the quality of both existing and newly created data has come to the forefront. Those in charge of managing these spatial datasets need the ability to control how they are created and edited to maintain this quality.

The geodatabase contains many parameters and options that the user can preset to help maintain data integrity in spatial datasets. Knowing what these components are and how to design them is crucial to building a new geodatabase, whether it is for local or enterprise use or even for use with online or mobile apps. And once these features are built into your data schema, it is important to know how to place your data into the geodatabase to take full advantage of these tools. Maintaining data integrity doesn't stop with building and importing data; it extends into editing existing data and creating new data. The editing interface includes many tools that work with the geodatabase to enforce the rules of data integrity, so the proper use of these tools is critical to maintaining your data at the highest standards. It's even possible to hide the nature of these data integrity rules so that they don't interfere with the normal workflow, but they can be used to steer the user toward the proper data entry techniques. This book explores many of those tools, including default values, field domains, subtype categories, relationship rules, editing constraints and data entry control, custom feature templates, and more—all of which, when combined, will provide a solid editing experience designed to maintain your data integrity.

This book will take you through designing, building, and editing your datasets using all the latest features and enhancements. Each chapter discusses the major concepts and includes hands-on tutorials that take you step by step through various focused topics. The tutorials include a sample project file in which you can practice the skills outlined, datasets to work with, special Your Turn segments in which you can practice earlier concepts, a review with study questions for classroom use, and an independent exercise that you can work through to ensure you have mastered the concepts. Chapter topics include the careful design of a geodatabase schema, building geodatabases that include data integrity rules, populating your geodatabases with existing data, sharing the data on the web and building 3D views, creating new features, editing the data with various techniques, and working with topologies.

Data for the book's tutorials is available in ArcGIS Online. For instructions on downloading it, see "Access the data" in chapter 1.

For a 180-day free trial of ArcGIS® Desktop software, go to the book's resource web page at esri.com/FocusGDBs, and use the EVA code inside the back cover of this book.

Instructor resources are available by request, at esripress@esri.com.

Chapter 1

Designing the geodatabase schema

One of the most important things you can start with in creating a geodatabase is a good design—and that design starts with a well-thought-out process about what data the geodatabase will house, how it will be edited, and how it will be maintained over time. Of course, a geodatabase will contain features classes, but whether to include a feature dataset, field names, domains, subtypes, and other functionalities should all be considered, and designed, before your fingers ever touch the keyboard. Case in point, this entire first chapter can be worked strictly with paper and pencil. These elements are all complex and cannot be created on the fly if you hope to achieve maximum success. Think about each of the components, examine what type of data will be used, imagine yourself or others having to edit this data, or even think about how you might use the data in a geoprocessing task. Perhaps you've used a poorly designed dataset before and spent a large amount of time formatting the data just to make it usable. If you can remember what problems arose and what steps you took to correct them, you will have a good understanding of how to create your datasets and avoid those mistakes.

It may also help to imagine what a final output map containing this data might look like so that you can determine what fields to use for labeling and symbolizing. The field types you design will make calculations either very easy or very difficult. Labeling will either be a snap or a headache. All these things can affect the design of a geodatabase, and a good design will make all future work that much easier.

Tutorial 1-1: Creating a geodatabase—building a logical model

The newest release of ArcGIS® Desktop software includes many additions and improvements to the data storage capabilities of the Esri® geodatabase file structure. There are new techniques to control interactions with the data, assign a behavior to it, and define relationships among datasets. In the design process, it is important to understand these techniques to build the most efficient database possible. It is even possible to build data controls to aid the user in creating data while decreasing the chance of introducing errors into the dataset. This book was tested using ArcGIS Pro 2.3.

LEARNING OBJECTIVES
- Outline geodatabase behavior
- Integrate datasets
- Model reality
- Use ArcGIS® Pro

Introduction

The goal of designing a geodatabase is to model the reality it is intended to represent. There are many characteristics, or behaviors, of the data that can be included in a geodatabase using various techniques. As the data modeler, it is your job to explore the capabilities of ArcGIS to make the most efficient and flexible database possible. The time spent at the start of a project designing the geodatabase will reap rewards later by making the data easier to use and edit and presenting a better representation of reality.

The first step is to study the reality that is about to be modeled. Look carefully and determine what features must be included in the geodatabase. In ArcGIS, everything is modeled as points, lines, and polygons, so realistic characteristics will need to be assigned to these pieces.

Next, look at how the data will be created. Will it be imported from another source, collected with field equipment, traced from aerial photos, drawn from survey data, or derived from some other process?

Finally, consider how this data will be used. Who will perform the edits, and what queries might it be expected to support in the future? Knowing many of the questions that the data will be used to answer will shape the geodatabase design.

With these things in mind, you can start to construct a logical data model. The model will diagram your process and allow for updates and changes before the final design is committed to a geodatabase. The design process uses a spreadsheet-based form to enter all the characteristics of the geodatabase, the feature datasets, the feature classes, and the data integrity rules of the feature classes such as domains and subtypes. Microsoft Excel® spreadsheet files included with the exercise data follow the creation process in ArcGIS Pro that will be used for basic diagramming.

The logical model is used to show what data types you will have (points, lines, or polygons), what tabular data will be included in the dataset, and the relationships between the tabular data and feature classes (if any). The model is also easily shared among your colleagues, so you can get several opinions on the design you are attempting.

Once a preliminary logical model is completed, it can be checked against many of the advanced features of the geodatabase, such as domains, subtypes, and relationships. Have the best tools been employed to ensure data integrity, ease of editing, and future expansion? You will also check the geodatabase against your idea of how the data should behave.

The result should be a well-thought-out geodatabase that is both efficient and a good representation of reality . . . well, at least as close as you can get using points, lines, and polygons.

You will start with a simple geodatabase, and then examine several advanced geodatabase options to see if better efficiency and more realistic behaviors can be achieved. The good news about any geodatabase design is that if it works, it's a success. Ten different people could design 10 different geodatabases for the same project, and they all could work quite well. The true test is how efficient a geodatabase is, how well it models reality, how well it maintains data integrity, how flexible it remains for future projects, and how easy it is to work with in editing and extracting information.

Designing the data

Scenario

The City of Oleander, Texas, population 60,000, has hired you as the top-gun geodatabase designer and wants an all-new database design for its parcel data. The new geodatabase will be used to designate each piece of property in the city—who owns it, its legal description, address, and more. You'll get information from the city planner to get the full idea of what's needed, and then create a diagram of your proposal. At this point, the geodatabase will not be constructed, only designed.

The city planner describes a dataset that would have a polygon for each piece of property in the city, whether it is platted or unplatted. It should have information about the legal description, the street address, and the current usage of each property.

Data

Because you are creating this geodatabase from scratch, there is no data to start with. You will need to print the geodatabase design forms from the exercise materials you download from ArcGIS® Online and use them to document the design process.

Tools used

- Geodatabase design forms

Begin the logical design for the geodatabase

The main component of this geodatabase will be polygons representing every piece of property in the city. Each piece of property is assigned certain data by the city. This data includes the subdivision name, block designation, lot designation, street address, and a land-use code, which shows how the land is being used.

As you know, the geodatabase is the framework in which other components are built. It may contain feature classes, tables, relationship classes, feature datasets, and many other components. You will design the geodatabase and its components using the geodatabase design forms provided in the files you download for this book.

Access the data

The data is stored in a book group named Focus on GDBs in ArcGIS Pro (Esri Press) in the Learn ArcGIS organization. You will log in and access the files for this tutorial, as well as for all the tutorials and corresponding exercises in this book as you need them.

1. Go to https://www.ArcGIS.com, and log in with an ArcGIS Online account.

2. On the Home tab, type Focus on GDBs in ArcGIS Pro in the Search box, and then click the Search for Groups entry in the drop-down list. (If no groups were found, turn off the option to "Only search in your organization.")

3. Click the link to open the Focus on GDBs in ArcGIS Pro (Esri Press) group and find the data, named FocusGDB. It consists of a zip file for each tutorial in the book.

4. Create a folder named EsriPress in a location where you want to store all the files, preferably on your drive C. Click the thumbnail and download the data for the first tutorial, saving it to your new folder (e.g., C:\EsriPress), rather than in the Documents library or on the desktop.

5. Extract the zip file. It will create a folder named Tutorial 1-1.

6. You can download and access these files as you need them for the other tutorials in this book.

Start building the model

1. Open a file explorer window and navigate to the location where you downloaded the tutorial 1-1 materials. Open the file GDB design forms.xlsx. Note: There is also a PDF version of this file in the same location.

2. Print all six pages of the geodatabase design forms (GDB feature classes, tables, domains, domains 2, subtypes, and relationships).

You will write out all your designs on these printed sheets, and you can print more sheets for corrections or expanded designs. Pages 1 and 2 will be used for these first few steps, but the other pages will be used in the other steps of this tutorial.

The geodatabase will need a name. It should reflect in general terms what will be stored in it.

3. **On the first line of page 1 (GDB feature classes), write the name** LandRecords **for the geodatabase name.**

The next line asks for a feature dataset name. A feature dataset is used to separate data into smaller subsets but is also important in grouping data for use in topology and various other advanced features. For now, leave the feature dataset name blank.

The next step is to start filling in the feature classes. So far, the city planner has described only one feature class, which will contain the polygons representing parcels and will include the fields he described.

4. **On the feature class portion of the worksheet, add a new feature class named** Parcels. **Note its type as POLY (for Polygon), and give it an alias of** Property Ownership.

Geodatabase design forms		ArcGIS Pro
Geodatabase name		LandRecords
Feature dataset name		
Feature classes:		
Type	Feature class name	Alias
POLY	Parcels	Property Ownership
	Type:	Indicate if this is a PoiNT, Line, or POLYgon feature class.
	Name:	Enter the name of your feature class.
	Alias:	Describe the contents of the feature class.

The alias is one of the first characteristics of the geodatabase that will be assigned. This alias will be shown in the Contents pane when the layer is added to a project; consequently, it also could be used in a legend. The alias should be very descriptive of the data to distinguish it from other datasets.

This feature class will have fields to store data, and these fields are recorded on the second design form. From the city planner's description, you can determine that the feature class will have fields for the subdivision name, block designation, lot designation, street address, and a land-use code. The content of the first three fields is self-explanatory. They will need to contain alphanumeric characters, so their field types will be Text.

5. **On the Tables worksheet (page 2), write the name of the new feature class on the first line. Under the field name, enter the first field as Sub_Name. Note its type as Text. Add another field for Blk as Text and Lot_No as Text.**

Tables worksheet							
Feature class or table name	Field name	Alias	Field type	Nulls (Y/N)	Domain name or subtype field (S) or (D)	Default value	Length
Parcels	Sub_Name		Text				
	Blk		Text				
	Lot_No		Text				

Simple so far, but there is other information to enter that will start impacting the future use of the data. The first is the field alias, which is another characteristic of the geodatabase. The alias is typically similar to the field name, with the important difference that it is allowed to have spaces in the text. This alias will be shown in many of the ArcGIS tools, the attribute table, any classification schemes, and many more places when the data is accessed. The field alias also should be descriptive of what data the field contains.

6. **Next to the field name Sub_Name, write the description** Subdivision Name **as the alias. Then write the alias** Block Designation **for Blk and** Lot Number **for Lot_No.**

Tables worksheet							
Feature class or table name	Field name	Alias	Field type	Nulls (Y/N)	Domain name or subtype field (S) or (D)	Default value	Length
Parcels	Sub_Name	Subdivision Name	Text				
	Blk	Block Designation	Text				
	Lot_No	Lot Number	Text				

Those notations have taken care of some of the fields, but there are more. The next is street address information. The address could be entered as a single field, but if you ever want to geocode against this dataset, it would be better to have each component of the address in a separate field. The common fields for geocoding are street prefix type, prefix direction, address number, street name, street type, suffix direction, and zip code. Fields such as city name or state name may be necessary if you are geocoding a broader region, but because all the features that this dataset will contain are specific to Oleander, you can leave them out. All the listed fields must be included in the table.

One interesting thing is that if the fields are given certain names that ArcGIS uses as a standard for address components, they will be filled in automatically when you make an address locator. An address locator is a special file that ArcGIS builds using your dataset that will allow addresses to be found easily when geocoding or using the Find tool. This address locator can also be used in routing and network applications. A list of preferred field names for each field is stored in the address locator style file, which was loaded when you installed ArcGIS Pro. You can open the file in the <install directory>\ArcGIS\Pro\Resources\Locators folder and view the list by searching for the phrase *Preferred Field Names*. You may add your own field names to the list as needed.

7. **Open a file explorer window. Navigate to the folder containing your ArcGIS installation (e.g., C:\Program Files\ArcGIS\Pro\Resources), and open the Locators folder. Scroll down to the file USAddress.lot.xml, right-click, and click Open with > Notepad.**

8. **Use the Find tool, or scroll down to the area labeled "Reference data style for Single House."**

```xml
<!--This section defines the reference data styles, including reference data tables, fields, -->
<!--preferred field names, reference to the mapping schema, and join clauses for the primary -->
<!--and alternate name tables if it applies. The name of the style is displayed on ArcGIS -->
<!--geocoding Select Address Locator dialog box.-->
<ref_data_styles>
   <!--Reference data style for Single House (Address Points)-->
   <ref_data_style>
      <name>Single House</name>
      <type>Address</type>
      <priority>6</priority>
      <desc>US Single House Addresses</desc>
```

9. **Within the section, scroll down to the area labeled Primary.House.**

```xml
<field_role name="Primary.House" required="true">
   <display_name>House Number</display_name>
   <preferred_name>PremiseNumber</preferred_name>
   <preferred_name>HOUSE</preferred_name>
   <preferred_name>HOUSENUM</preferred_name>
   <preferred_name>HOUSE_NUM</preferred_name>
   <preferred_name>HOUSE-NUM</preferred_name>
   <preferred_name>HOUSE_NUMBER</preferred_name>
   <preferred_name>HOUSE-NUMBER</preferred_name>
   <preferred_name>HOUSE_NUMB</preferred_name>
   <preferred_name>HOUSE-NUMB</preferred_name>
   <preferred_name>HOUSENO</preferred_name>
   <preferred_name>HN</preferred_name>
   <preferred_name>NUMBER</preferred_name>
   <preferred_name>ADDNUMBER</preferred_name>
   <preferred_name>ADD_NUMBER</preferred_name>
   <preferred_name>ADD</preferred_name>
   <preferred_name>ADD_HN</preferred_name>
   <preferred_name>ADDRESS</preferred_name>
   <preferred_name>ADDRESS_NUM</preferred_name>
   <preferred_name>ADDR</preferred_name>
   <preferred_name>ADDR_HN</preferred_name>
   <preferred_name>SITENUMBER</preferred_name>
   <preferred_name>SITE_HOUSENUM</preferred_name>
   <preferred_name>SITUS_NUM</preferred_name>
   <preferred_name>STNUMBER</preferred_name>
   <preferred_name>ST_NUMBER</preferred_name>
   <preferred_name>STR_NUMBER</preferred_name>
   <preferred_name>STREET_NUM</preferred_name>
   <preferred_name>CIVICNUMBER</preferred_name>
</field_role>
```

As you can see, there are many acceptable names for this field. The advantage of using one of these suggested field names over a new incarnation is that prebuild geocoders and geoprocessing tools will automatically recognize these field names and, in some cases, can validate that you have the correct field name chosen for a given data type. Look over the rest of the list to see what the choices are for other field types such as Primary.StreetName, Primary.Locality, or Primary.Postal. It is important to try to use one of the preset abbreviations or add your own field names to this list whenever possible. This use of existing names will make the entry of field names in certain tool parameters almost automatic. The ones shown in this geodatabase design were all derived from this list.

10. **On the Tables worksheet, write the following field names, field types, and aliases:**
    - Pre_Type, Text, Prefix Type
    - Pre_Dir, Text, Prefix Direction
    - House_Num, Text, House Number
    - Street_Name, Text, Street Name
    - Street_Type, Text, Street Type
    - Suffix_Dir, Text, Suffix Direction
    - ZIP_Code, LI (Long Integer), ZIP Code

These fields will add a lot of functionality to the dataset that may be valuable later. For example, you could select all the parcels in a certain subdivision, all the parcels that front a certain street, or use the House_Num field to put address labels on the map.

The last bit of data that the city planner mentioned was the land-use code. The land-use code is split into two sets of codes. The primary land-use code is one of seven main codes, and each primary code has many secondary codes that are more descriptive of the land use. The use of two codes helps in making both a generalized land-use map and a detailed land-use map.

11. **On the Tables worksheet, add the field** Primary_Use, **with the field type** Text **and the alias** Primary Land Use Code, **and the field** Secondary_Code, **with the field type** Text **and the alias** Secondary Land Use Code.

The data entered so far has involved information that the city planner wanted. One more piece of data is necessary for you to maintain a connection to certain third-party data that is important to the project. The identity of the property owner is not stored in the parcel's attribute table but is stored in an external table. You will need to add a field to your data structure that will allow you to set up a relationship between the field and the external table. The procedure is discussed later in the tutorial, but for now, you will need to add a field to accommodate the relationship.

12. **On the Tables worksheet, add the field** Georeference, **with the field type** Text **and the alias** Georeference Index.

Tables worksheet

Feature class or table name	Field name	Alias	Field type	Nulls (Y/N)	Domain name or subtype field (S) or (D)	Default value	Length
Parcels	Sub_Name	Subdivision Name	Text				
	Blk	Block Designation	Text				
	Lot_No	Lot Number	Text				
	Pre_Type	Prefix Type	Text				
	Pre_Dir	Prefix Direction	Text				
	House_Num	House Number	Text				
	Street_Name	Street Name	Text				
	Street_Type	Street Type	Text				
	Suffix_Dir	Suffix Direction	Text				
	ZIP_Code	ZIP Code	LI				
	Primary_Use	Primary Land Use Code	Text				
	Secondary_Code	Secondary Land Use Code	Text				
	Georeference	Georeference Index	Text				

Design for data integrity

The design looks good so far, but imagine what will happen when people start putting data in the table. If they left the Sub_Name field blank, there would be no way to identify the legal record of a piece of property. What about address number or the land-use code? These fields shouldn't be left blank, or there could be gaps in the data. On the other hand, not every street will have a value for prefix type, so there will be instances when a field value can be left blank and still be correct. You can set up parameters in the geodatabase to control these data integrity rules.

One way to build data integrity rules into your table is to set the flag for allowing null values, or no value, for a field. If nulls are not allowed, a validation check of the data will produce an error for any records entered without all the necessary values being provided. Perhaps, as an example, the person entering the data accidentally skipped the field during data entry or tried to enter data before all the information was known. Either way, it could cause problems with your data if nulls are not allowed.

The solution is to mark in the design table which fields are allowed to have nulls and which must have a value entered.

1. **On the Tables worksheet, mark the following fields to allow null values by placing a** Y **in the Nulls column:**
   - Pre_Type
   - Pre_Dir
   - Suffix_Dir

2. Mark the remaining fields as not allowing null values by placing an *N* in the Nulls column:
   - Sub_Name
   - Blk
   - Lot_No
   - House_Num
   - Street_Name
   - Street_Type
   - ZIP_Code
   - Primary_Use
   - Secondary_Code
   - Georeference

Feature class or table name	Field name	Alias	Field type	Nulls (Y/N)
Parcels	Sub_Name	Subdivision Name	Text	N
	Blk	Block Designation	Text	N
	Lot_No	Lot Number	Text	N
	Pre_Type	Prefix Type	Text	Y
	Pre_Dir	Prefix Direction	Text	Y
	House_Num	House Number	Text	N
	Street_Name	Street Name	Text	N
	Street_Type	Street Type	Text	N
	Suffix_Dir	Suffix Direction	Text	Y
	ZIP_Code	ZIP Code	LI	N
	Primary_Use	Primary Land Use Code	Text	N
	Secondary_Code	Secondary Land Use Code	Text	N
	Georeference	Georeference Index	Text	N

Another data integrity component is the domain. A domain allows you to define a list of values for any text field or a range of values for a numeric field. When data is entered, it is matched against the domain to see if it is a valid value. This designation helps eliminate typos or inventive abbreviations. Imagine 10 data entry clerks all coming up with unique abbreviations for the land-use code Vacant. It might be entered as VAC, V, Vcnt, or any number of misspellings. A query to find all vacant property would be difficult. If a domain is applied to the field Primary_Use that contains only the seven correct category abbreviations, it would be impossible for anyone to enter a value that wasn't in the domain.

In addition to the defined primary-use codes, there are secondary-use codes associated with each primary-use code. For instance, the primary code Commercial has a set of secondary codes named Light Commercial, Special District, Church, and School. Using a contingent value domain, you can restrict the field selections for a second field on the basis of the first field's value. The Secondary Land Use Code field will get a domain with all the available land-use descriptions in it, but later you will build a matrix that will pair the Primary Land Use Code with only those values in the domain that are relevant to the selected code. This pairing will prevent the user from selecting a secondary-use code that does not match the category of the primary-use code.

The domain values will be entered on the Domains worksheet, and you will note in the worksheet that it is a domain to avoid confusion with subtype fields that may be entered later.

3. On the Tables worksheet, add an entry on the line for Primary_Use with the name of a domain that will contain the acceptable values for this field. Call it Prim_Use_Codes, and place a (D1) in front of it for "domain number 1."

4. On the line for Secondary_Code, add the notation (D2) and the name Sec_Use_Code.

| Primary_Use | Primary Land Use Code | Text | N | (D1) Prim_Use_Codes |
| Secondary_Code | Secondary Land Use Code | Text | N | (D2) Sec_Use_Code |

5. Now turn to the Domains worksheet (page 3), and write the domain name Prim_Use_Codes, with a description of Primary Use Codes for Parcels, a field type of Text, and the type of domain as Coded Values.

6. In the Code column, write VAC, with a description in the Value column of Vacant Property. Under that, write RES, with a description of Residential Property. Continue down the form, entering the rest of the Prim_Use_Codes values from the accompanying list. Print more worksheets if necessary.

    COM   Commercial Property
    IND   Industrial Property
    GOV   Government Property
    PRK   Park Land
    OTHER Other Uses

Domains worksheet

	Domain name	Description	Field type	Domain type	Coded values Code (Min)	/ Range Value (Max)
D1	Prim_Use_Codes	Primary use Codes for Parcels	Text	Coded Values	VAC	Vacant Property
					RES	Residential Property
					COM	Commercial Property
					IND	Industrial Property
					GOV	Government Property
					PRK	Park Land
					OTHER	Other Uses

Adding this domain will build a validation check for data integrity. You can rest assured that the primary-use code abbreviation entered for any piece of property will fit your normal list. But one concern might be that someone could set the primary-use code to Commercial, and then set the secondary-use code to 35th3. These values don't match, so a domain for the secondary codes should be designed to include all the detailed codes. In the form, a note can be added to show which primary land-use code each value will pair with, and later you will see how these codes are paired to further constrain the values entered.

7. **On the Domains worksheet, add a new domain name (D2) of** Sec_Use_Code. **Give it a description of** Secondary Land Use Codes, **with a field type of** Text **and a domain type of** Coded Values. **The codes that will go in this domain are as shown in the figure. Print more copies of the worksheet as needed.**

Coded values		Range	
Code	(Min)	Value	(Max)
VAC	PRIV	Vacant Private Land	
VAC	GOV	Vacant Government Land	
RES	SF_DET	Single Family Detached	
RES	SF_LIM	Single Family Limited	
RES	DUP	Duplex	
RES	TRI	Triplex	
RES	QUAD	Quadruplex	
RES	CONDO	Condominiums	
RES	MANUF	Manufactured Housing	
RES	TOWN	Townhomes	
RES	MULTI	Multi-Family	
COM	LT_COM	Light Commercial	
COM	SPEC	Special District	
COM	SCH	School	
COM	CHR	Church	
IND	LT_IND	Light Industrial	
IND	HVY_IND	Heavy Industrial	
GOV	CITY	City Owned	
GOV	CNTY	County Owned	
GOV	STATE	State Owned	
GOV	US	Federally Owned	
PRK	PUB	Public Park	
PRK	PRIVPRK	Private Open Space	
OTHER	ROW	Right-of-way	
OTHER	PROW	Private Right-of-way	
OTHER	UTIL	Utility	
OTHER	ESMT	Easement	

Perhaps there are other fields in the table that would benefit from the application of a domain. Most of them, however, such as a subdivision name or a house number, couldn't be constrained in this way; there would be too many values. But the field street type might be a good candidate. The US Postal Service has a standard set of street type abbreviations, and from time to time, you may be asked to generate a mailing list from this table. So, it would be a good idea to add a domain to this field.

The table includes many acceptable street type abbreviations; you would not want to list all of them on the Domains form, and you wouldn't want to type them into a domain. So, a command exists to take a file listing of street types and read them into a domain, and a file named Suffix.txt with these abbreviations is provided in the downloaded materials. The abbreviations it contains were found on a US Postal Service website, and the file contains all the recognized suffix names. The process of turning this file into a domain will be demonstrated in tutorial 2-1, but for now you can write the file name Suffix.txt on the design worksheet.

8. **On the Tables worksheet, write** St_Type_Abbrv **as the domain name for the field Street_Type, and add** (D3), **noting that it is a domain.**

9. **Next, go to the Domains worksheet, and write the name** St_Type_Abbrv, **a description of** Street Type Abbreviations, **a field type of** Text, **and a domain type of** Coded Values. **Under Code, write the file name to identify the file holding the domain values.**

| D3 | St_Type_Abbrv | Street Type Abbreviations | Text | Coded Values | Tutorial 1-1/Suffix.txt |

Now is a good time to investigate other aspects of how the data will be used and see if there are any other data integrity techniques that might be employed—most notably, the subtypes.

Consider the situation with property. It is either platted by a legal survey or unplatted and recorded as a single deed. You might separate property as either platted (divided into developed lots with utilities) or unplatted (raw agricultural land). It is important to know the distinction for legal purposes and for the sale of property. It would be possible to put the platted land into one feature class and the unplatted property in another. If both feature classes were in the same geodatabase, they could both be easily stored and edited at the same time. The symbology and annotation would work well for both, and each feature class could have different data integrity rules. So dividing them into two feature classes would work and might be beneficial.

But consider how the data might be used in a query. If a list of all property owners in a given region was needed, it would have to come from two separate files, and exporting the list would create two tables. Although it would be beneficial in some respects to put property data into two feature classes, using the data would be problematic. That's where subtypes can be valuable.

Using a subtype is a way to create a virtual subdivision of data within the same feature class, and then apply different data integrity rules to each category. It's the best of both worlds: the data can be separated into logical categories and be given data integrity rules for each category but keep the convenience of being edited, queried, and managed in a single feature class. You'll also see later, in tutorial 1-2, how subtypes can be used to set default values, establish unique attribute domains, set connectivity rules, and establish relationship rules for each subcategory created. They'll even make it easier to symbolize and label data.

A field to contain the subtype code must be added to the table. The field type must be Integer, and the codes will be established along with a description. For this data, you'll make a code 1 for Platted Property, code 2 for Unplatted Property, and code 3 for Plat Pending. This last code will be for property that has been approved by the city but is awaiting the filing data from the county. This will be a simple subtype, without any additional data integrity rules added.

10. On the Tables worksheet, add a new field on the bottom named Plat_Status, **make its field type** SI (short integer), give it an alias of Plat Status, and don't allow for null values. Because most new property being added to the dataset will be platted, record its default value as 1. Finally, write the name Plat_Subtype **for the subtype name, with a notation of** (S1).

| Plat_Status | Plat Status | SI | N | (S1) Plat_Subtype |

11. Go to the Subtypes worksheet (page 5). Write the name of the subtype as Plat_Subtype, **and add the three codes described previously:**
    1 = Platted Property
    2 = Unplatted Property
    3 = Plat Pending

Subtypes worksheet				PRESET DEFAULTS	
Subtype name	Code	Description	Field	Domain name	Default value
S1 Plat_Subtype	1	Platted Property			
	2	Unplatted Property			
	3	Plat Pending			

Extend the data model

This work concludes the initial design phase of the Parcels feature class, but there's another component to investigate. When these polygons are symbolized, they can each have a solid fill and a line style for their perimeter. When maps are made, however, the boundaries of the parcels must be symbolized differently. The edge of the parcel that fronts a street will be drawn with a thicker line; the edges representing property lines between properties will be a thinner line; if someone owns two adjacent pieces of property, the line between them should be dashed.

Consider creating a set of lines that will duplicate the boundaries of each parcel. Then these lines can be symbolized as described. The only field the feature class will need is a code describing which type of line to draw. This field would benefit from having a data integrity rule (a domain) with the three categories of lines described.

A behavior will need to be created between the polygons representing property and the lines representing their boundaries. If the shape of any polygon is modified, the lines will need to automatically adjust to coincide. This type of relationship is called *topology* and will be discussed in chapter 7. For ArcGIS to manage this topology, the feature classes must reside within the same feature dataset.

Feature datasets are another way to segregate data inside a geodatabase. If any behavior is to be built for a feature class, such as topologies, network databases, geometric networks, relationships, or terrains, the feature class must reside in a feature dataset. For this example, you will establish a feature dataset for your feature classes, so that the corresponding topology can be built.

1. **On the Geodatabase worksheet (page 1), write the name of the feature dataset as** PropertyData. **Next, write the new feature class name** LotBoundaries **on a blank line. Give it a feature type of** LINE **and an alias of** Lot Boundaries.

Geodatabase design forms		ArcGIS Pro	
Geodatabase name		LandRecords	
Feature dataset name		PropertyData	
Feature classes:			
Type	Feature class name	Alias	
POLY	Parcels	Property Ownership	
LINE	LotBoundaries	Lot Boundaries	

2. **Next, go to the Tables worksheet, and write the name of the new table as** LotBoundaries. **Then write the single attribute of this table, Line_Code. Give it a field type of** Text, **add an alias of** Line Code, **and do not allow nulls. Add a notation that there is a domain for this field, and name it** (D4) Parcel_Line_Codes.

LotBoundaries	Line_Code	Line Code	Text	N	(D4) Parcel_Line_Codes

3. **Finish by filling in the information for the domain. On the Domains worksheet (page 3), add the name of the domain as** Parcel_Line_Codes, **a description of** Line Codes for Parcels, **and a field type of** Text, **and note the domain type as** Coded Values. **Then write the three domain values described previously:**
   - ROW = Edge of Right-of-way
   - LOT = Lot Line
   - SPLIT = Split Lot Line

D4	Parcel_Line_Codes	Line Codes for Parcels	Text	Coded Values	ROW	Edge of Right-of-way
					LOT	Lot Line
					SPLIT	Split Lot Line

Design a relationship class

The features you've dealt with in the design so far have been the points, lines, and polygons that will create the model of reality. Not all the data you will need for this model, however, is in the form of points, lines, and polygons. The design will also need to include tabular data that is provided by an outside agency. For each parcel, a county appraisal agency provides ownership and value information. This data would be valuable for analysis if it was associated with the parcel data. The nature of the table is that it is updated regularly from separate appraisal software, so it cannot be incorporated in the polygon feature class in the same way as regular data. By keeping it separate, it will facilitate the maintenance of both ArcGIS use of the data and the third-party software's use of the data.

A relationship class has many of the benefits of a simple join in a project but also provides a mechanism for controlling edits in the related table. If the graphic features are altered in editing, rules in the relationship class can also alter the related table and maintain the relationship. For this example, the parcels have a match in the appraisal roll table. If a piece of property is removed because of replatting, the associated record in the appraisal table can be set to be deleted automatically.

The final consideration is the cardinality of the relationship. If each parcel has one and only one match in the appraisal table, and vice versa, the cardinality is said to be one to one (1:1). If one parcel can have several matches in the appraisal table, such as the case of a single parcel being owned by more than one person, the cardinality is said to be one to many (1:M). If the opposite relationship was also true—that is, an owner can also own several pieces of property—the relationship is said to be many to many (M:N).

Armed with this information, you can move to the worksheet on relationship classes and fill in the details.

Relationship worksheet

Name of the relationship class:	
Origin table/feature class:	
Destination table/feature class:	
Relationship type:	Simple (peer to peer)　　　Composite
Labels:	
Origin to destination:	
Destination to origin:	
Message propagation:	Forward　　Backward　Both　　None
Cardinality:	1-1　　　1-M　　M-N
Attributes:	No　　　Yes - Table name:

Add to the tables worksheet

	Primary key field	Foreign key name
Origin table/feature class:		:
Destination table/feature class:		:

1. Continuing in the same design spreadsheet, on the Relationships worksheet (page 6), write the origin table as Parcels and the destination table as TaxRecords_2019. Name the output relationship class Ownership.

Relationship worksheet

Origin table/feature class:	**Parcels**
Destination table/feature class:	**TaxRecords_2019**
Output relationship class name:	**Ownership**

The relationship class can be used to add or delete records, but because the related table will be managed by another source, the relationship type should not allow records to be deleted, making it a simple (peer-to-peer) relationship. Labels will be shown to describe the relationship between the tables. The description for moving from the parcels feature class to the appraisal table is "Parcel is owned by," and from the appraisal table to the parcels feature class, it is "Owner has ownership of." As the relationship is used in analysis, these labels will remind the user of the nature of the relationship. Normally, relationship classes are transparent to the user, but you can have your project controls display a message when the relationship is used. For this example, opt not to use them.

2. **Circle Simple (peer to peer) as the relationship type, and write the labels** Parcel is owned by **for Forward Path Label and** Owner has ownership of **for Backward Path Label. Circle None for Message propagation.**

Next, you'll note the cardinality as many to many, since a parcel can be owned by several people, and one person may own several parcels. It may also be beneficial to store what percentage of ownership can be attributed to each owner. This notation will help when more than one person is recorded as the owner. You'll write the name of the table as Ownership_Rel, and it will be added to the Tables worksheet later. Finally, you'll select the fields that will be the basis for the relationship and give them a label describing their relationship to the related table (foreign key name).

3. **On your Relationship worksheet, circle M-N for Cardinality, and circle Yes under Attributes. Set the origin table and destination table primary key fields as** Georeference. **Name the origin table foreign key** Owner **and the destination table foreign key** Property.

Relationship worksheet				
Origin table/feature class:	**Parcels**			
Destination table/feature class:	**TaxRecords_2019**			
Output relationship class name:	**Ownership**			
Relationship type:	Simple (peer to peer)		Composite	
Labels:				
Forward Path Label:	**Parcel is owned by**			
Backward Path Label:	**Owner has ownership of**			
Message propagation:	Forward	Backward	Both	None
Cardinality:	1-1	One to one		
	1-M	One to many		
	M-N	Many to many		
Attributes:	No	Yes		
		Primary key field		Foreign key name
Origin table/feature class:	**Georeference**	:	**Owner**	
Destination table/feature class:	**Georeference**	:	**Property**	

This work completes the logical model for the geodatabase. From these design forms, you will be able to create the entire structure and begin using it for storing data. If you do not have a lot of experience editing geodatabases, you may want to jump ahead to tutorial 2-1 and see how this design will function, and then come back to this exercise. Otherwise, complete this exercise, which will continue to focus on the design phase.

Exercise 1-1

The tutorial showed how to diagram a geodatabase to include feature classes along with their associated tables, domains, and subtypes. The goal was to think through the design, adding data integrity and behavior guidelines to the database.

In this exercise, you will repeat the process for another dataset required by the city planner. This one will contain the zoning data for Oleander. The zoning code for a piece of property determines the type of development that is allowed on a parcel (even though the land use may be different). The zoning districts may incorporate several parcels and generally follow parcel boundaries, but they can split parcels, too.

The zoning districts will be represented by solid shaded polygons, so you will want to design a polygon feature class for districts. The edges of the polygons should be symbolized in one of two ways—either as a solid line representing a zoning boundary or a dashed line representing a change in allowable development density. Because of this scenario, you will want to design an additional linear feature class for symbology purposes. The codes necessary for the zoning information are as follows:

Code	Description
R-1	Single Family Residential
R-1A	Single Family Attached
R-1L	Single Family Limited
R-2	Duplex
R-3	Triplex
R-4	Quadruplex
R-5	Multifamily
C-1	Light Commercial
C-2	Heavy Commercial
TH	Townhomes
LI	Limited Industrial
I-1	Light Industrial
I-2	Heavy Industrial
TX-121	121 Development District
POS	Public Open Space

Analyze the descriptions of this data and determine what feature datasets and feature classes must be made, what fields they should contain, any domains that might

need to be created (and possibly contingent domains), and any subtypes that might be beneficial.
- Print a set of the geodatabase design forms as necessary.
- Use the forms to create the logical model for feature classes for the zoning polygons and zoning boundaries.
- Investigate the use of domains and subtypes to build data integrity and behavior into your design.

WHAT TO TURN IN

If you are working in a classroom setting with an instructor, you may be required to submit the design forms you created in tutorial 1-1.
- The completed geodatabase worksheets for:
    - Tutorial 1-1
    - Exercise 1-1

Review

Over the last 30 years, the way that geographic features have been portrayed, stored, and manipulated in geographic information systems (GIS) has evolved from a file-based technology into the present-day Esri geodatabase format. By using the Esri geodatabase, GIS practitioners can more realistically manage geographic features and their relationships to other features. Although computer technology has enhanced the behavioral aspects of these relationships, the fundamental ways that these geographic features are represented—by points, lines, and polygons—has largely remained unchanged. Esri geodatabase technology has improved the management of these points, lines, and polygons by providing tools to create geographic feature representations, enforce data integrity, and establish relationships among the geographic features that more closely model real-world situations.

As illustrated in the previous exercise, the opportunities to manage data using GIS methodology can be enhanced by careful thought and preplanning to ensure that an accurate portrayal of geographic features and their relationships is contained in the geodatabase. Preplanning the geodatabase is enhanced through a structured, organized logical data model to ensure that every conceivable relationship is accounted for in the model. This preplanning phase is no easy task. However, it is much easier to spend time at the outset designing your geodatabase than it is to change it once you've begun entering data into the model.

Organizing your geodatabase using feature classes and feature datasets allows you to refine relationships and behaviors for the data. Feature classes, as the most basic representation of geographic data in the geodatabase, can be logically grouped together to form feature datasets. Although there are many different techniques for organizing geographic data in the geodatabase, the organization of the data must be guided by the behavior of these features in the real world. For example, if feature classes contained in the geodatabase work together to form a geometric network, represent a terrain, or establish a topology, the feature classes must reside in the same feature dataset. Such behaviors among the data must be considered while designing the geodatabase.

Once your design is complete, using domains for your attribute data and other techniques will reduce costly mistakes during the data entry phase of your geodatabase's development. Additional techniques provided by the geodatabase, such as the creation of subtypes, optimize how data is organized and utilized within the geodatabase. Using the many tools available within your project, and with a thoroughly planned approach, your new geodatabase will adequately portray the geographic features and associated relationships among them. As a result, your model of reality as contained in the geodatabase will represent the real-world features as closely as possible.

STUDY QUESTIONS

Answers to the study questions in this book are available on the instructor resources DVD.
1. What is a logical model of a geodatabase, and why should you develop a logical data model when designing your geodatabase?
2. What are the principal advantages of using subtypes? Give one example of a situation in which you would create a subtype, and specify why.
3. What are the principal advantages of using domains?
4. What is the difference between feature classes and feature datasets? When must you create a feature dataset?

Other study topics

Search for these key phrases in ArcGIS Pro Help for further reading:
1. Fundamentals of the geodatabase
2. What is a geodatabase?
3. Introduction to attribute domains
4. Fields, domains, and subtypes

Tutorial 1-2: Creating a geodatabase–expanding the logical model

The components of a geodatabase can have various spatial relationships, or behaviors, that form a topology. These behaviors can exist among points, lines, and polygons and will impact the logical model of a database. The most efficient designs will consider topology from the beginning.

LEARNING OBJECTIVES
- Design linear feature classes
- Investigate data behavior
- Design for topology
- Design point feature classes

Introduction

The first tutorial used the geodatabase design forms to construct a logical design for a parcels database. That dataset consisted of a polygon feature class along with a linear feature class to aid in symbolizing the parcel boundaries.

In this tutorial, you will design another set of feature classes to store data for a sewer system. The process will include investigating the behavior of the data, and then trying to accommodate it in the design.

Remember to look at how the data will interact with feature classes, as well as any possible domains or subtypes that may be used. This investigation will help to build not only an efficient design but also a good model of reality.

Designing the data structure

Scenario

After your successes with the parcels and zoning datasets, Oleander's Public Works Department is seeking your help to create a geodatabase for the sewer system. You will need to design this geodatabase for them.

Sewer systems are a simple design. They consist of pipes to carry wastewater to the treatment plant. In real life, it's important that the pipes connect to ensure a direct flow route from the beginning of the system to the end. In the data model, you will also want to ensure connectivity, which will allow the data to be used later to construct a network dataset.

A great amount of data can also be stored as attributes of the linear features. Some of the basic information includes the size of the pipe, the material it's made of, and the year of installation.

This portion of the process is intended to inspire thought and creativity. If done correctly, your designs will be viable for years to come. Once all the designs are completed, they will be used to create the data structure in ArcGIS Pro.

Data

Since you are creating this geodatabase from scratch, there is no data to start with. But you will need to print the geodatabase design forms from the downloaded materials as an aid in the design process. Print as many of the pages as necessary to contain all your designs.

Tools used
- Geodatabase design forms

Begin the geodatabase design process

Using the geodatabase design forms, you will once again commit your thoughts to paper and examine all aspects of how the data will be used, edited, and symbolized. The first part of the design will be to name the geodatabase. Since the data likely will be used in a network later, it will also require a feature dataset.

1. **On the first page of a new set of design forms, write the name of the new geodatabase as** Utility_Data. **On the next line, add the name of the feature dataset as** Wastewater.

   The sewer lines will be built as linear features, which will require a feature class. Several attributes can be stored with the lines, as mentioned earlier. You'll add each of these attributes to the design forms.

2. **On the geodatabase design form, write the name of the new feature class,** SewerLines, **with a feature type of** LINE **and an alias of** Sewer Lines.

Geodatabase design forms		ArcGIS Pro
Geodatabase name		Utility_Data
Feature dataset name		Wastewater
Feature classes:		
Type	Feature class name	Alias
LINE	SewerLines	Sewer Lines

   Next, you will need to fill in the Tables worksheet and show which fields the feature class will contain. The three fields that were required by Public Works were pipe size, which can be a number; pipe material, which can be text; and the year the pipe was installed, which is also a number.

3. **On the Tables worksheet, write the name of the feature class. Then write the fields** Pipe_Size **with a data type of** SI, Material **with a data type of** Text, **and** Year_Built **with a data type of** LI. **Add the aliases of** Pipe Size, Pipe Material, **and** Year Built, **respectively.**

   In Oleander, some of the sewer lines that run through the city belong to other agencies. Some are the pipes of other cities and are headed for the treatment plant, and some belong to the regional utility that handles all the wastewater treatment for local cities. They all must be included in the dataset and on the maps to prevent accidentally digging into them.

The owner of the line must also be recorded, so you'll add a field named Description to store the name of the owner of each pipe.

4. **In the Field Name column, add a field named** Description. **Write a field type of** Text **with an alias of** Owner.

Tables worksheet							
Feature class or table name	Field name	Alias	Field type	Nulls (Y/N)	Domain name or subtype field (S) or (D)	Default value	Length
SewerLines	Pipe_Size	Pipe Size	SI				
	Material	Pipe Material	Text				
	Year_Built	Year Built	LI				
	Description	Owner	Text				

Data integrity issues

For this data, it is important that every pipe have an entry for size and material. However, year of construction may not be known for some of the older, existing pipes. So, do not accept null values for the fields Pipe_Size and Material, but allow nulls for Year_Built. Also, the ownership of every pipe must be known, so don't allow for nulls there.

1. **For each field, write** N **next to the aliases Pipe Size, Pipe Material, and Owner in the Nulls column. Write** Y **next to Year Built in the same column.**

The next data integrity issue is to investigate the use of domains. Sewer pipes vary in size from 6 inches to 12 inches in 2-inch increments. Pipes larger than 12 inches are called *interceptors* and are metered to determine the charge to the city. In Oleander, the interceptors are owned by a regional utility that handles all the wastewater processing. Although the pipes run through the city, Oleander's Public Works Department does no service or maintenance on them.

If a domain was built for pipe size, it could prevent some data entry errors. The choices would be to use coded values and enter a discrete set of values or use a range and give a low and a high value, such as 6 and 12. A range would allow any numeric entry between these values, and because the sizes increase in 2-inch increments, there would be values allowed by the domain that are not allowed in reality. For example, the range from 6 to 12 would allow an entry of 9, but there is no such thing as a 9-inch sewer pipe. So, using a range wouldn't work. It is apparent that a discrete list of coded values should be entered.

2. **On the Tables worksheet, write the name of the domain for the field Pipe_Size as** Sewer_Pipe_Size **with the (D1) notation. Then on the Domains worksheet, write the same name. Add a description of** Sewer Pipe Size, **set the field type as SI, and write the domain type as** Coded Values. **Enter the values as shown and their corresponding descriptions:**
   - 6 = 6"
   - 8 = 8"

- 10 = 10"
- 12 = 12"

Domains worksheet						
					Coded values / Range	
Domain name	Description		Field type	Domain type	Code (Min)	Value (Max)
D1 Sewer_Pipe_Size	Sewer Pipe Size		SI	Coded Values	6	6"
					8	8"
					10	10"
					12	12"

Notice that although the field stores integers, and the code must be an integer, the associated description can be text. The description will be useful in labeling the text later, as the inch marks will be visible on the labels that ArcGIS Pro generates.

Another data integrity tool is to include subtypes. Subtypes can be used to segregate data, so that there will be separate domains and defaults for each subset of data. In this scenario, the data might be separated by material. Almost all the new PVC pipes going in are 8 inches, almost all the new high-density polyethylene pipes are 10 inches, and almost all the ductile iron pipes going in are 12 inches. These three are the only materials allowed for new pipes, so if each of these materials was set up as a subtype, additional control could be added to automatically populate some of the more common fields.

One problem with this approach would be the interceptors. These pipes are typically larger than 12 inches, but the size and material change for each situation. Default values wouldn't be appropriate here, so the interceptors don't play by the same rules as the Oleander pipes. Perhaps the solution is to put them in their own feature class. They could still be edited simultaneously with the Oleander data, and they could still participate in any networks that are built, as long as they reside in the same feature dataset. Also, the fields for interceptors would be identical to the Oleander lines. You'll update the worksheets to include an additional feature class for interceptors.

3. **On the geodatabase worksheet, add the name of the new linear feature class as** Interceptors, **and give it an alias of** Interceptors. **Be sure to fill in its type as** LINE.

4. **Write the name of the feature class on the Tables worksheet, and duplicate all the fields from the SewerLines feature class.**

Because the interceptors don't have any regular size or material, there will be no domains or default values for these lines. With the problem solved, you can proceed to design the subtypes. A good field that could use a subtype is Material. By selecting the material, the default values will automatically populate the other fields. And if a pipe size other than the standard is used, the pipe size domain will prevent any incorrect values from being entered.

The subtype field must always be an integer, and the material field is set as text. This entry can be changed easily with an eraser.

5. On the Tables worksheet, erase the field type for Material and enter SI. Also, add the name Sewer_Line_Material in the Subtype column on the right with an (S1) notation.

Field name	Alias	Field type	Nulls (Y/N)	Domain name or subtype field (S) or (D)	Default value	Length
Material	Pipe Material	SI	N	(S1) Sewer_Line_Material		

Next, you can fill in the Subtypes worksheet for the first material, polyvinyl chloride (PVC). The default value for PipeSize will be 8 inches, and the domain designed previously should be applied to this field. The default for the description field will be Oleander. And to save a little typing, make the default for Year_Built 2010. You can change it once a year to keep up with new construction.

6. On the Subtypes worksheet, write the name of the Subtype field, Sewer_Line_Material. Write the code as 1 and the description as PVC. In the Field column, write the name Pipe_Size, note its domain as Sewer_Pipe_Size, and its default value as 8.

Note that the default value does not have the inch marks next to it. Remember that its field type is short integer, so the value entered in the database must be short integer. But the inch mark is stored in the domain description, which can be used for labeling later if necessary. The benefit here is that you can do math processes on the pipe size value, such as a selection PipeSize ≤ 8, but still have the value with an inch mark for labeling.

7. Continuing on the Subtypes worksheet, write the names of the other fields and their default values. For the Description field, write a default value of Oleander. For the Year_Built field, write a default value of 2019.

This work completes the design for the first choice of subtype. The next choice will be for the material type of high-density polyethylene (HDPE). The default size will be 10 inches, and the defaults for description and year built will be the same as before.

8. On the next blank line of the Subtypes worksheet, write the code of 2 with a description of HDPE. In the Field column, write the name Pipe_Size, note its domain as Sewer_Pipe_Size, and its default value as 10. As before, write a default value of Oleander for the Description field and 2019 as the default value for the Year_Built field.

Subtypes worksheet

Subtype name		Code	Description	Field	PRESET DEFAULTS	
					Domain name	Default value
S1	Sewer_Line_Material	1	PVC	Pipe_Size	Sewer_Pipe_Size	8
				Description		Oleander
				Year_Built		2019
		2	HDPE	Pipe_Size	Sewer_Pipe_Size	10
				Description		Oleander
				Year_Built		2019

YOUR TURN

Fill in the information for a third subtype code with the material type ductile iron, or DI. It will have a default size of 12 inches with the same domain as the other sizes, as well as a default description of Oleander and default year built of 2019.

There are two more material types, and although they are no longer installed new, they could cause validation problems later if they are not included in the design. The two types are concrete and clay. You'll add them as choices 4 and 5 on the Subtypes worksheet. They will require no domains or defaults because they will not be used to enter new pipes.

9. On the next blank line, write a code 4 with a description of Conc and a code 5 with a description of Clay. No defaults or domains are required for these subtypes.

Subtypes worksheet				Field	PRESET DEFAULTS Domain name	Default value
	Subtype name	Code	Description			
S1	Sewer_Line_Material	1	PVC	Pipe_Size	Sewer_Pipe_Size	8
				Description		Oleander
				Year_Built		2019
		2	HDPE	Pipe_Size	Sewer_Pipe_Size	10
				Description		Oleander
				Year_Built		2019
		3	DI	Pipe_Size	Sewer_Pipe_Size	12
				Description		Oleander
				Year_Built		2019
		4	Conc			
		5	Clay			

This work completes the design for linear features. Next will be investigating the point features associated with the sewer lines. At each intersection of sewer lines, and at various locations along their length, manholes are constructed for maintenance. At the ends of the lines, a smaller access port called a *cleanout* is added to accommodate the mechanical device that is run down the pipes to clean out clogs. The cleanouts will be represented in the geodatabase by points, with certain attributes associated with them.

These points have a behavior relationship with the lines, in that they must fall on top of the lines. If any networking is done, the points must be snapped to the lines to preserve connectivity. They must also reside in the same feature dataset.

The data associated with the points will include several fields. The first will be a code that marks which points are manholes and which are cleanouts. Other information such as the flow line, rim elevation, and depth (rim elevation minus the flow line elevation) will be copied from the construction documents. Finally, a field for the year of construction and another for description (ownership) will be needed.

10. **On the geodatabase design form, write the name of the new point feature class as** SewerFixtures. **Give it a feature class type of** PNT **(Point) and a description of** Sewer Fixtures.

Next, add the fields for the point feature class on the Tables worksheet.

11. **On the Tables worksheet, write the name of the new feature class, and add the following field names with their field types, aliases, and null value allowances:**
    - Fix_Type, Fixture Type, SI, No
    - Flowline, Flowline Elevation, Float, Yes
    - Rim_Elev, Rim Elevation, Float, Yes
    - Depth, Depth from Surface, Float, Yes
    - Year_Built, Year Built, LI, Yes
    - Description, Owner, Text, No

Even though there won't be defaults or domains for any of these fields, it might be useful to make fixture type a subtype. One benefit of subtypes is that each code in the subtype list can be selected as a target when editing. Without a subtype, you would set the target as SewerFixtures and click to add a point. Then you would have to set the fixture type immediately because it cannot be null. With a subtype set for fixture, the target drop-down list would show the two types of fixtures allowed. As each new point is entered, the fixture type field is automatically populated, meaning that it can never be null. For the short amount of time it takes to set up the subtype structure, it would be a great way to enforce the data integrity rule of not allowing null values. At the same time, a default for description and year built could be added for convenience. You'll start by noting the subtype name on the tables form, and then populate the subtype form.

12. **On the Tables worksheet, write the subtype name** Sewer_Fix_Type **for the Fix_Type field with a notation of** (S2). **Under SewerFixtures, for the following fields, add a default Year Built value of** 2019 **and a default Description value of** Oleander.

Tables worksheet							
Feature class or table name	Field name	Alias	Field type	Nulls (Y/N)	Domain name or subtype field (S) or (D)	Default value	Length
SewerFixtures	Fix_Type	Fixture Type	SI	N	(S2) Sewer_Fix_Type		
	Flowline	Flowline Elevation	Float	Y			
	Rim_Elev	Rim Elevation	Float	Y			
	Depth	Depth from Surface	Float	Y			
	Year_Built	Year Built	LI	Y		2019	
	Description	Owner	Text	N		Oleander	

13. **On the Subtypes worksheet, write the name of the subtype as** Sewer_Fix_Type. **Give it a code** 1 **for** Manhole **and a code** 2 **for** Cleanout.

    The interceptors will also have associated fixtures, but they are all manholes. This factor makes for a simple feature class because all the features will be symbolized the same. They will have the same fields as the Oleander fixtures, except for fixture type and description. These fields are unnecessary because their values would always be the same.

14. **Add a new feature class to the Geodatabase worksheet (page 1). Name it** InterceptorFix, **with a feature type of** PNT **and a description of** Interceptor Fixtures.

15. **Fill in the Tables worksheet for the new feature class** InterceptorFix **with these fields and values:**
    - Flowline, Flowline Elevation, Float, Yes
    - Rim_Elev, Rim Elevation, Float, Yes
    - Depth, Depth from Surface, Float, Yes
    - Year_Built, Year Built, LI, Yes **(default value of** 2019**)**

Feature class or table name	Field name	Alias	Field type	Nulls (Y/N)	Domain name or subtype field (S) or (D)	Default value	Length
InterceptorFix	Flowline	Flowline Elevation	Float	Y			
	Rim_Elev	Rim Elevation	Float	Y			
	Depth	Depth from Surface	Float	Y			
	Year_Built	Year Built	LI	Y		2019	

The geodatabase design forms make this design seem simple, but it is a fairly complex database. A good deal of thought was put into the fields required for the feature classes, the relationships of the feature classes, and the inclusion of data integrity rules, such as defaults, domains, and subtypes.

Review the design forms and resolve any questions that you may have, because the next tutorial, 2-1, will have you build these data structures in ArcGIS Pro.

Exercise 1-2

The tutorial showed how to apply design strategies and data integrity rules to point and linear feature classes. Each feature type was analyzed against the reality it is supposed to model to build as much behavior and data integrity as possible.

In this exercise, you will repeat the process with storm drain data. Oleander's Public Works Department would like a geodatabase design for the storm drain system, just like the one you did for the sewer collection system. It will consist of the pipelines and fixtures associated with them. The lines are all made from reinforced concrete pipe (RCCP) and vary in size from 15 inches to 45 inches in 3-inch increments. The pipes are usually classified as laterals (21 inches or less), mains (longer than 21 inches), and boxes (square pipe with no restriction in size). The data for these features includes a pipe size, material, description, flowline in, flowline out, slope, year installed, and a designation for a public or private line.

Connected to these pipes are various types of fixtures listed as follows. These fixtures will be used as the subtypes, and their code from the existing data is included:

101 = curb inlet
102 = grate inlet
104 = junction box
106 = Y inlet
107 = junction box/manhole
108 = outfall
109 = headwall
110 = beehive inlet
111 = manhole

The type of data collected for these features includes a description, flowline elevation, inlet size, top elevation, year built, designation for a public or private line, and a rotation angle.

Analyze the descriptions of this data and determine what feature datasets and feature classes need to be created, what fields they should contain, any domains that might need to be created, and any subtypes that might be beneficial.

- Print a set of the geodatabase design forms as necessary.
- Use the forms to create the logical model of feature classes for the zoning polygons and zoning boundaries.
- Investigate the use of domains and subtypes to build data integrity and behavior into your design.

Getting started

Here's a little help to get started:
- Decide how many feature classes you want to create.
- List the fields that will need to be in each feature class.
- Determine the field type, null status, and default value for each field.
- Investigate the use of domains for these fields.
- Look for fields that describe a "type" or "category" that could be used as a subtype, such as fixture type.

WHAT TO TURN IN

If you are working in a classroom setting with an instructor, you may be required to submit the design forms you created in tutorial 1-2.
- The completed design worksheets for:
  - Tutorial 1-2
  - Exercise 1-2

Review

Whereas the first tutorial focused on the development of a geodatabase to contain parcel-related information, this tutorial focused on the development of a geodatabase to represent a sewer system. Both tutorials focus on real-world examples of the types of critical information managed by every city, town, or other local governmental entity. Given the tremendous municipal resources dedicated to the management of these systems, an adequate foundation for managing them is essential. The geodatabase allows for physical information, as well as information on the behaviors among the components of these systems, to be accurately portrayed in a GIS. By using the combination of good information and good data integrity controls (behavior), the geodatabase enhances effective and efficient decision-making capabilities.

One of the most critical steps in developing a comprehensive geodatabase is the initial planning phase. Although we have focused on and discussed GIS functionality within both tutorials, we have yet to even start a create process in the software! Planning for a geodatabase development project involves both gathering the physical system requirements and understanding how interrelated objects behave in the physical system. To further enhance your design, you will also need to know what kinds of questions your customers will need to have answered. Once an adequate knowledge of the system is gained, it is then up to the skill of the geodatabase developer to build these capabilities into the corresponding geodatabase.

A properly designed geodatabase will be worthless to the engineers and others who rely on it to represent the real world if it is full of errors. The geodatabase allows the designer to apply *domains* to feature attributes to ensure that the correct information is correctly recorded within the geodatabase. Flexibility to adequately portray and control features and their behavioral characteristics within the model is afforded using *subtypes* of features. Interrelated behavior among features is enhanced using *topology*. The success of a geodatabase often depends on a thorough knowledge of how and when to apply these data integrity tools.

STUDY QUESTIONS

1. What is topology, and why is it an important concept when designing a logical model of a geodatabase?
2. Think about the principal ways in which features are represented spatially within GIS. Give an example of each feature type, a possible domain for each type, and possible subtypes for each type. Explain your rationale.
3. Why is it important to fully understand a system to be represented and managed in GIS? How do you determine its structure?

Other study topics

Search for these key phrases in ArcGIS Pro Help for further reading:
1. Introduction to subtypes
2. Geoprocessing considerations for attribute domains
3. Geoprocessing considerations for subtypes

Chapter 2

Creating a geodatabase

Chapter 1 covered many aspects of designing a geodatabase, and now it's time to build one.
A slow, methodical approach is best, making sure that each detail of your planned design is implemented exactly. In this chapter, you will create geodatabases, feature datasets, and feature classes—and apply all the domains and subtypes that were introduced in chapter 1. Then at the end of this chapter, you will get a chance to create and edit some data within this new structure and see how a good design makes data creation and management easier.

Tutorial 2-1: Building a geodatabase

Designing a geodatabase can be a long and drawn-out process, but it's only after that phase is completed that the data structure can be created in ArcGIS Pro. A good design will greatly simplify the creation phase and make it go smoother and faster.

LEARNING OBJECTIVES
- Work with ArcGIS Pro
- Create a geodatabase
- Build a database schema

Introduction

A good database design is essential for the smooth creation of a geodatabase. It is best to think through the entire design, documenting your needs and addressing them with the logical model. Once that is completed, the creation phase can begin.

Databases can be sensitive to change, and in fact, some elements, once created, cannot be changed. They can only be deleted, and then re-created. It is therefore important when using the creation tools to pay attention to what has been designed. Creating a feature class without regard to the data type will result in wasted time because the data type cannot be altered later. The same holds true for field names and null allowances. Once created, these things cannot be changed.

The order of creating components is not necessarily as critical. Feature classes created outside a feature dataset can usually be moved into a feature dataset with a simple drag-and-drop operation. Field names left out of tables can usually be added later, or subtypes can be set up after the fact.

But even though some alterations can be made to the data structure later, it is best to prepare a complete logical model and follow it closely when creating the geodatabase. At this point, all opinions should have been heard and all aspects of the design completed. Major alterations to a data schema after the fact may require exporting the data to new feature classes, which could result in errors and confusion.

Creating the data structure

Scenario

The City of Oleander has passed along completed geodatabase design forms and asked that you build the data structure for them. The forms show feature datasets, feature classes, fields, domains, and subtypes. They have some existing data that will be loaded into this schema later.

Data

Print out the design forms provided in the downloaded materials for tutorial 2-1. These are the completed geodatabase logical model diagrams from tutorials 1-1 and 1-2.

Tools used

ArcGIS Pro:
- New file geodatabase
- New feature dataset
- New feature class
- Create a domain
- Table To Domain

As you go through the creation process, pay close attention to all the steps involved. Sometimes setting even one option wrong can result in having to delete the entire piece and start over. The order of creation is not as critical. Some people may prefer to make all the domains first, then create all the tables, and follow that up with subtype creation. Others will create one feature class, along with its table, domains, and subtypes, and then move on to the next table. If the components are understood and created correctly, the order is not important.

Create the data structure

You are starting with a blank canvas and will be creating everything from scratch. The first step will be to create the geodatabase, which will contain the rest of the components.

Chapter 2 | Creating a geodatabase 35

This could be an enterprise geodatabase or file geodatabase, depending on your setup, but for ease and clarity, this tutorial will create a file geodatabase.

1. **Open a file explorer and navigate to the EsriPress folder. The completed designs for this tutorial can be found in the Tutorial 2-1 folder, but you may use your own completed forms from tutorial 1-1.**

2. **Print the geodatabase design forms from tutorial 1-1.**

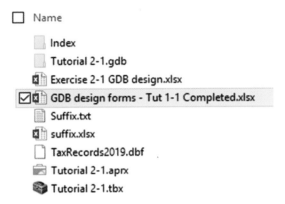

Each ArcGIS Pro project contains a local geodatabase. This geodatabase can be used to store all the data associated with a project, or you can store your data elsewhere and use this geodatabase for temporary files or files that would not be used with other projects. In this instance, you will create a new geodatabase in the project where you will build all the files for your design.

3. **In the Catalog pane, right-click Databases and click New File Geodatabase.**

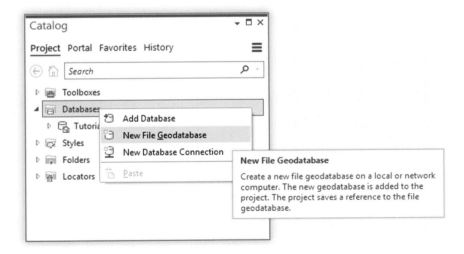

4. **Provide the name** LandRecords **as shown in your design forms, and click Save.**

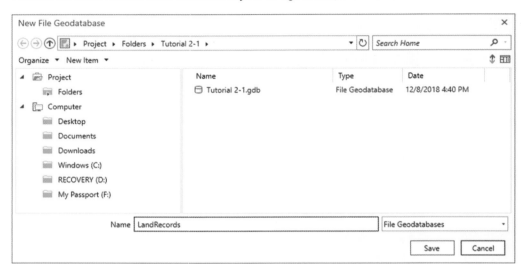

Create a feature dataset

From the design form, you can see that there will be a feature dataset in this geodatabase, with two feature classes. The geodatabase itself only needed a name to exist. The feature dataset needs two things: a name and a spatial reference.

The spatial reference is what ties your data to a location on the globe. You may need to decide whether your data should be stored as projected or unprojected data, and you will need to define the spatial extent of your dataset. This dataset must cover the City of Oleander in north central Texas and will use a spatial reference that is typical for the area. For more information on selecting the spatial reference for other datasets, see the ArcGIS Pro Help topic "An overview of spatial references."

1. **In the Catalog pane, expand the Databases folder, and then right-click LandRecords and click New > Feature Dataset.**

2. **A Geoprocessing pane will open for the Create Feature Dataset tool. Type the name of the database that's on the design form of** PropertyData.

3. **To set the coordinate system, click the Globe icon to open the selection dialog box. Click Projected Coordinate System > State Plane > NAD 1983 (US Feet). Scroll down the list, and select NAD 1983 StatePlaneTexas N Central FIPS 4202. Click OK. Hint: Search for** FIPS 4202, **and expand Projected coordinate system, State Plane, and NAD 1983 (US Feet) to reveal this entry as the only choice. You may also want to click Add to Favorites to access this coordinate system even faster in the future.**

Chapter 2 | Creating a geodatabase   37

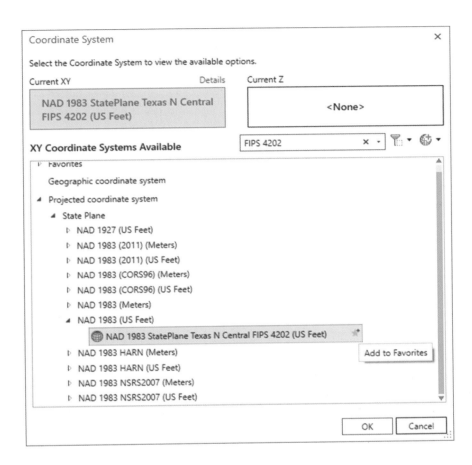

4. Click Run to complete the process. When completed, click the Catalog tab at the bottom to return to the Catalog pane for the next step.

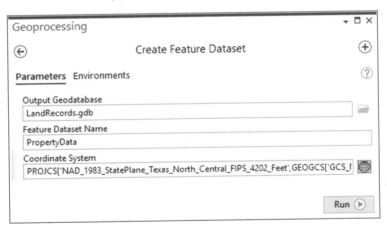

Create the feature classes

Inside this feature dataset, you will create two feature classes. A feature class, at the minimum, needs three things: a name, a spatial reference, and a geometry type. You will provide the name from the design form. The spatial reference has been set at the feature dataset level, and the feature class will inherit this spatial reference. The final thing will be to set the geometry type. Note: Be sure to set this type correctly, as it cannot be changed later.

1. **Right-click the PropertyData feature dataset. Click New > Feature Class. The tool dialog box will open in the Geoprocessing pane.**

2. **Type the name and alias as shown on the design worksheet. Then click the data type drop-down arrow and click Polygon.**

   Note that this is page 1 of 6. The dialog box for creating the new feature class has five other screens, which can be accessed by clicking the Next button. In this first example, you can skip them, but they will be demonstrated later in this chapter.

3. **Click Finish, and when completed, close the Geoprocessing pane.**

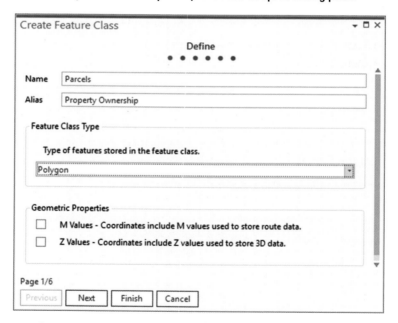

For this first feature class you create, you will use the Catalog pane to build the fields, but in the future, you can create the fields in the same process as creating the feature class.

The groundwork for the feature class has been created. Next is to enter all the fields where the attributes will be stored. Refer to the Tables worksheet for the names and configurations of the fields, but at this point you should disregard the settings for Domain and Default (those settings will be configured later). As you enter each field, check the worksheet to get all the settings correct. Then highlight the row and double-check all the entries before moving to the next field. This attention to detail will help you avoid mistakes, which could cause you to have to delete the feature class and start over.

Feature class or table name	Field name	Alias	Field type	Nulls (Y/N)
Parcels	Sub_Name	Subdivision Name	Text	N
	Blk	Block Designation	Text	N
	Lot_No	Lot Number	Text	N
	Pre_Type	Prefix Type	Text	Y
	Pre_Dir	Prefix Direction	Text	Y
	House_Num	House Number	Text	N
	Street_Name	Street Name	Text	N
	Street_Type	Street Type	Text	N
	Suffix_Dir	Suffix Direction	Text	Y
	ZIP_Code	ZIP Code	LI	N
	Primary_Use	Primary Land Use Code	Text	N
	Secondary_Code	Secondary Land Use Code	Text	N
	Georeference	Georeference Index	Text	N
	Plat_Status	Plat Status	SI	N

4. **Right-click the new feature class, and click Design > Fields.**

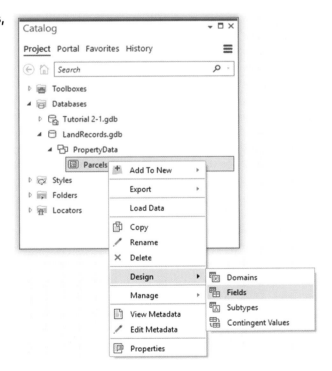

5. Notice that the ribbon menu is changed to reflect tools associated with feature class design.

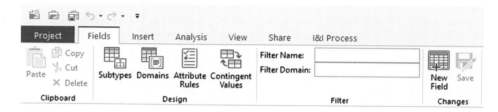

6. Click the bottom line that reads, "Click here to add a new field." Type the field name Sub_Name, the alias Subdivision Name, and set the data type to Text. Click Enter after each entry. Click to clear the check box under Allow NULL. Hint: Remember that you will configure the defaults and domains later.

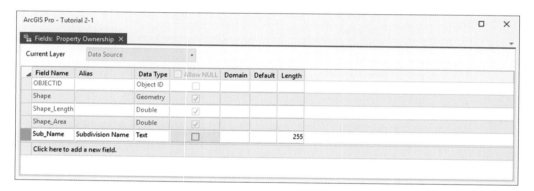

Be careful and methodical as you work your way through the field entry process. It is desirable to do it in one pass through the dialog box to avoid getting confused or forgetting to set a value. Once you press the Finish button, only the alias can be changed later. Mistakes must be deleted and the whole file, or table, constructed again.

7. Click to add another field. On the blank line, type the field name Blk, enter the alias as Block Designation, and set the field type to Text. Click to clear the Allow NULL check box.

YOUR TURN
Work your way down the list of fields from the Tables worksheet, and enter all the values for Field Name, Alias, Data Type, and Allow NULL columns, disregarding the Default and Domain/Subtype entries. If you run out of blank lines for new fields, use the slider bar on the right to reveal more.

When you have finished entering all the fields, click Save on the Field tab, in the Changes group. Close the Fields creation pane.

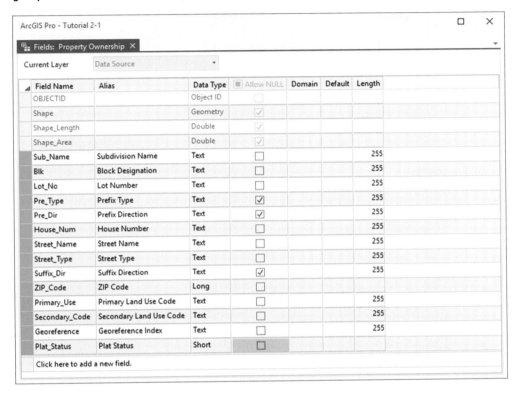

The feature class has been created and added to the feature dataset. There's one more feature class to create for the lot boundaries. This feature class has only one attribute field, so you can speed up the process by adding the fields to the feature class in the primary feature class creation screen.

8. In the Catalog pane, right-click the PropertyData feature dataset. Click New > Feature Class.

9. Type the name and alias as shown on the design form. Then click the data type drop-down arrow and select Line. Click Next.

10. As with the other type of field creation screen, click the last line to add a new field.

11. **Type the field name and parameters from the design worksheet. Click Finish.**

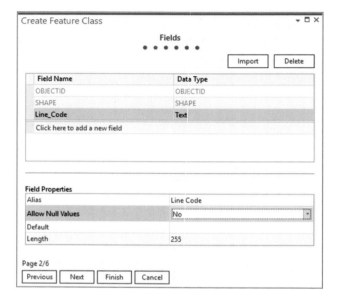

Note that the feature class with fields was created in one process. Although this is simple for a feature class with just a couple of fields, more complex feature class creation will benefit from using the spreadsheet-based parameter entry screens.

Create the domains

Now you have both feature classes, with all their fields. The next step is to introduce data integrity rules for the feature classes. This process starts with creating the domains, and then assigning them to the fields in the feature classes. Domains are stored at the geodatabase level, which means that all the required domains for all the feature classes residing in the geodatabase will be stored there. You may be working on a utility project, but there may be other feature classes requiring dozens of domains. Because of this possibility, there are two critical rules when working with domains.

The first rule is to give your domain a specific name, even if it means making it long. A domain named Type may be confusing, because it doesn't describe its purpose well. Is this a type of roads, signs, trees, or something else? A better name might be Type_Of_Signpost if it was used to constrain the entry of signposts to certain types, such as metal or wood.

The second rule is to never alter someone else's domain to fit your needs. You may find a domain named Material, with a list of wood, steel, fiberglass, and concrete. The domain you want may be the same list but without concrete, so you delete that entry and use it. The next time the dataset for which that domain was originally created is used, all the features with a material of concrete will be invalid with respect to the Material domain because the concrete entry is missing. If it gets changed back to make that data valid, your new dataset will become invalid when referencing that domain. To prevent this problem from happening, it's best to always create your own domains.

You will start with the Primary Land Use Codes domain. The Domains worksheet will be the guide for creating domains, so get it ready.

	Domain name	Description	Field type	Domain type	Code (Min)	Value (Max)
D1	Prim_Use_Codes	Primary Use Codes for Parcels	Text	Coded Values	VAC	Vacant Property
					RES	Residential Property
					COM	Commercial Property
					IND	Industrial Property
					GOV	Government Property
					PRK	Park Land
					OTHER	Other Uses

Remember that there are two types of domains: coded values and range. The first, coded values, contains a list of values. When you set the field type to Text, the only choice is coded values, so it is set automatically. Numeric fields can also be used with the Coded Values choice. The second domain, range, lets you set a start and an end value within which the entered data must fall, but it is valid only with numeric field types.

1. In the Catalog pane, right-click the geodatabase LandRecords, and then click Domains. In the Domains pane, enter the name of the domain, the description, the field type, and the domain type.

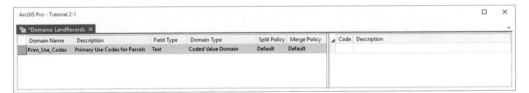

The right side of the Domains pane has a matrix for entering the codes and descriptions, so now focus your actions in that area.

2. On the first line in the Code column, enter the first value of VAC. Next to it in the Description column, enter the value Vacant Property. Click Save on the Domains tab, in the Changes group, but leave the Domains pane open.

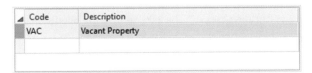

YOUR TURN

Using the Domains worksheet, enter the rest of the values for code and description. Periodically, click Save to save your work, and make sure there are no errors in the domain. When you have completed the list, click Save one final time. Hint: Use the Enter key to move between the entries. If the box next to the domain name turns red when you save, click it to identify the error (most commonly having an empty code on the last line). When the error is fixed, click the red error box and try saving the domain again.

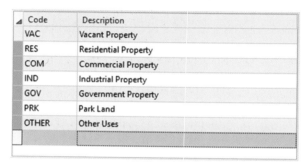

Next is the domain for Secondary Land Use Codes. Even though this domain will take some extra time to create, it is an important part of the data integrity rules and will pay future benefits.

Coded values	/	Range
Code (Min)	Value	(Max)
VAC	PRIV	Vacant Private Land
VAC	GOV	Vacant Government Land
RES	SF_DET	Single Family Detached
RES	SF_LIM	Single Family Limited
RES	DUP	Duplex
RES	TRI	Triplex
RES	QUAD	Quadruplex
RES	CONDO	Condominiums
RES	MANUF	Manufactured Housing
RES	TOWN	Townhomes
RES	MULTI	Multi-Family
COM	LT_COM	Light Commercial
COM	SPEC	Special District
COM	SCH	School
COM	CHR	Church
IND	LT_IND	Light Industrial
IND	HVY_IND	Heavy Industrial
GOV	CITY	City Owned
GOV	CNTY	County Owned
GOV	STATE	State Owned
GOV	US	Federally Owned
PRK	PUB	Public Park
PRK	PRIVPRK	Private Open Space
OTHER	ROW	Right-of-way
OTHER	PROW	Private Right-of-way
OTHER	UTIL	Utility
OTHER	ESMT	Easement

Follow the diagram carefully to get all these domains entered, and save when completed.

Code	Description
PRIV	Vacant Private Land
GOV	Vacant Government Land
SF_DET	Single Family Detached
SF_LIM	Single Family Limited
DUP	Duplex
TRI	Triplex
QUAD	Quadruplex
CONDO	Condominium
MANUF	Manufactued Housing
TOWN	Townhouse
MULTI	Multi-Family
LT_COM	Light Commercial
SPEC	Special District
SCH	School
CHR	Church
LT_IND	Light Industrial
HVY_IND	Heavy Industrial
CITY	City Owned
CNTY	County Owned
STATE	State Owned
US	Federally Owned
PUB	Public Park
PRIVPRK	Private Open Space
ROW	Right-of-way
PROW	Private Right-of-way
UTIL	Utility
ESMT	Easement

There are two more domains to create. One will use the same process as the use codes domain, and the other will use a process to import the values from an existing database. You'll create the parcel line codes domain first because it uses the same dialog box as the previous example, but you will start the process differently.

3. **On the Domains tab, in the Changes group, click New Domain. A new line will be added to the Domains pane. Click the first empty line, and type the new domain name** Parcel_Line_Codes **and the description** Line Codes for Parcels.

46 Focus on Geodatabases in ArcGIS Pro

4. **Set the field type to** Text. **The domain type will automatically set to coded values. In the Code pane, enter the codes and descriptions for the domain:**
   - ROW: Edge of Right-of-way
   - LOT: Lot Line
   - SPLIT: Split Lot Line

   When all the values are entered, click Save in the Changes group. When you're finished, close the Domains pane.

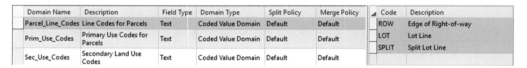

Create a domain from a table

The next method of domain creation uses an existing database of street suffixes to fill in all the values. ArcGIS Pro has a tool that will read the table and transfer it into the domain you created in the geodatabase. The first step is to find the tool.

1. **On the ribbon, click the Analysis tab, and in the Geoprocessing group, click Tools to open the Geoprocessing pane.**

2. **On the Find Tools line, type** Table To Domain.

3. Click the Table To Domain tool to open it.

4. In the tool dialog box, click the Browse button next to Input Table. Browse through the C:\EsriPress\FocusGDB\Projects\Tutorial 2-1 folder for the table Suffix.txt. Select the table, and click OK.

5. In the Code Field input box, click the drop-down arrow and select SuffixAbbrv.

6. In the Description Field input box, select SuffixType.

7. For Input Workspace, use the Browse button to set it to the LandRecords geodatabase, and click OK.

8. **Type the domain name as** St_Type_Abbrv **and the domain description as** Street Type Abbreviations. **When your dialog box matches the figure, click Run. When it completes, close the Geoprocessing pane.**

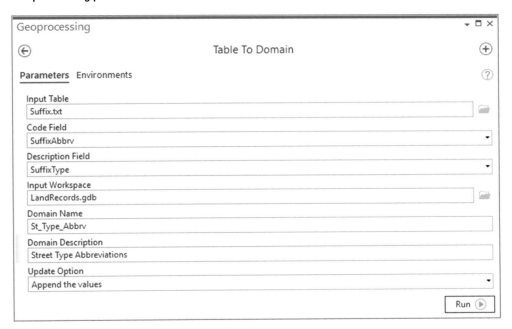

The domain was created with all the suffix values that are used by the US Postal Service. This is a quick, easy option for creating domains that come from other data sources.

9. Open the Domains pane of the geodatabase LandRecords and examine the results. Close the Domains pane.

Assign the domains

Now that all the domains are created, they need to be assigned to the fields that they will help control. The Tables worksheet shows that these domains will be assigned to three fields in the Parcels feature class and one field in the LotBoundaries feature class.

Parcels	Sub_Name	Subdivision Name	Text	N		
	Blk	Block Designation	Text	N		
	Lot_No	Lot Number	Text	N		
	Pre_Type	Prefix Type	Text	Y		
	Pre_Dir	Prefix Direction	Text	Y		
	House_Num	House Number	Text	N		
	Street_Name	Street Name	Text	N		
	Street_Type	Street Type	Text	N	(D3) St_Type_Abbrev	
	Suffix_Dir	Suffix Direction	Text	Y		
	ZIP_Code	ZIP Code	LI	N		
	Primary_Use	Primary Land Use Code	Text	N	(D1) Prim_Use_Codes	
	Secondary_Code	Secondary Land Use Code	Text	N	(D2) Sec_Use_Code	
	Georeference	Georeference Index	Text	N		
	Plat_Status	Plat Status	SI	N	(S1) Plat_Subtype	
LotBoundaries	Line_Code	Line Code	Text	N	(D4) Parcel_Line_Codes	

1. In the Catalog pane, right-click the Parcels feature class, and click Design > Fields to open the Fields pane.

2. In the Fields pane, find the field Street_Type, and double-click the blank space in the Domain column. From the drop-down list, click St_Type_Abbrv. Click Save on the Fields tab, in the Changes group.

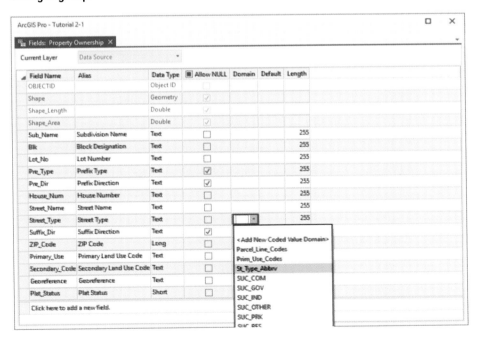

Notice that when you select the domain, the drop-down list is by name and not description. This listing is another reason to give domains explicit names instead of something less descriptive or generic such as Material or Type.

3. **Find the field Primary_Use, and double-click the space in the Domain column. Select Prim_Use_Codes in the drop-down list. Click Save.**

Next, you will need to set up the contingent domains for the Secondary Land Use Code field. This listing is done by first applying the Secondary Land Use Codes domain, and then building the Contingent Values field group.

4. **Apply the Sec_Use_Code domain to the Secondary_Code field. Click Save and close the Fields pane.**

5. **In the Catalog pane, right-click the Parcels feature class, and click Design > Contingent Values.**

No field groups will exist, so you must make a new one from scratch.

6. **Click the box to add a new field group.**

7. **Type the name** Secondary Use Codes. **Click Add Fields, and click the boxes next to PrimaryUse and SecondaryCode. Then click Add and OK to finish.**

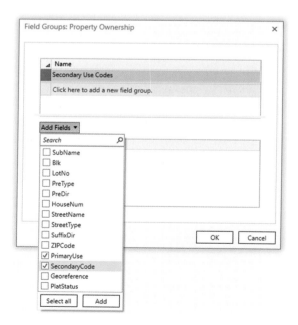

8. Click on the first line to add a new contingent value.

9. Use the drop-down boxes for each field to set the values to Vacant Property and Vacant Private Land. Click Save.

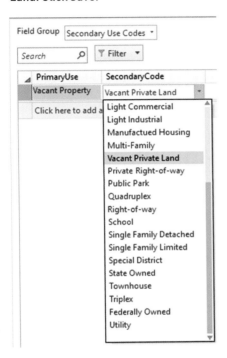

YOUR TURN

Follow the Domains worksheet shown at the top of this section, under "Assign the domains," and add the value pairs to the field group matrix. Once again, this will be a long and tedious process, but it will pay many future benefits when these data integrity rules help keep your data error free. When you are finished adding these pairs, click Save.

PrimaryUse	SecondaryCode
Vacant Property	Vacant Private Land
Vacant Property	Vacant Government Land
Residential Property	Single Family Detached
Residential Property	Single Family Limited
Residential Property	Duplex
Residential Property	Triplex
Residential Property	Quadruplex
Residential Property	Condominium
Residential Property	Manufactued Housing
Residential Property	Townhouse
Residential Property	Multi-Family
Commercial Property	Light Commercial
Commercial Property	Special District
Commercial Property	School
Commercial Property	Church
Industrial Property	Light Industrial
Industrial Property	Heavy Industrial
Government Property	City Owned
Government Property	County Owned
Government Property	State Owned
Government Property	Federally Owned
Park Land	Public Park
Park Land	Private Open Space
Other Uses	Right-of-way
Other Uses	Private Right-of-way
Other Uses	Utility
Other Uses	Easement

There's only one more domain to build, and you should be able to finish the rest on your own.

YOUR TURN

Apply the domain Parcel_Line_Codes to the Line_Code field in the LotBoundaries feature class, starting with step 1 in this section. Remember that domains must be applied to the fields before they have any effect on controlling your data integrity.

Set up subtypes

As recognized earlier, having subtypes in the Parcels feature class can be a benefit. A subtype can show the plat status and allow updates to be done more efficiently. If you are still unsure of the role of subtypes, the next section will demonstrate how they can be used to great advantage.

Refer to the Subtypes worksheet for the information that will be included in the subtype entry screen.

	Subtype name	Code	Description
S1	Plat_Subtype	1	Platted Property
		2	Unplatted Property
		3	Plat Pending

1. In the Catalog pane, right-click the feature class Parcels, and click Design > Subtypes. Next, on the Subtypes tab, in the Subtypes group, click Create/Manage. Notice that the Subtypes tab on the ribbon was selected automatically.

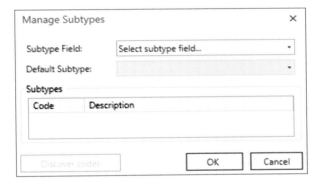

2. For Subtype Field, use the drop-down list to set it to Plat_Status.

3. Set the first code to 1, and type the description of Platted Property. Then fill in the remaining entries according to the Subtypes worksheet:
   - 2 = Unplatted Property
   - 3 = Plat Pending

4. When your dialog box matches the figure, click OK.

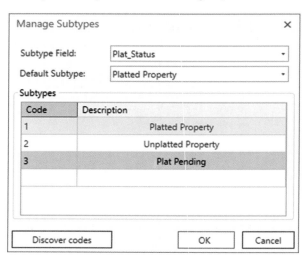

5. View the Subtypes matrix to see the results, and then click Save and close the Subtypes and Fields panes.

In this example, there were no additional default or domain values to set unique to the subtype setting. Pressing Save was a good way to check if there were any problems with your entries. If an error had been encountered, the line containing the error would have turned red and instructed you how to correct the error and proceed.

DATA INTEGRITY ON THE WEB
You have successfully designed, built, and implemented a set of domains and subtypes. If you later want to host this data as a feature service in ArcGIS Online or Portal for ArcGIS®, these domains and subtypes will be honored. Setting these parameters means that setting values through a web interface will show valid lists from your domain and help ensure data integrity. The subtypes will be active as well, but only if the layers are symbolized using the Unique Values classification using the Subtype field as the symbol field.

One word of caution is that the structure of these items cannot be changed in the feature service. If you later want to add another selection to your domain or another class to your subtype, you cannot. That makes it even more important to spend the time designing and documenting your databases before creating and using them.

Define the relationship class

You designed a relationship class in chapter 1, in tutorial 1-1, that will link the parcel polygons to an external table of ownership data. This relationship class will be built in the geodatabase for land records, and it can access only data stored within this geodatabase. Review the design form for relationship classes before proceeding.

Relationship worksheet

Origin table/feature class:	Parcels		
Destination table/feature class:	TaxRecords_2019		
Output relationship class name:	Ownership		
Relationship type:	**Simple (peer to peer)**	Composite	
Labels:			
Forward Path Label:	Parcel is owned by		
Backward Path Label:	Owner has ownership of		
Message propagation:	Forward	Backward	Both	**None**
Cardinality:	1-1	One to one	
	1-M	One to many	
	M-N	Many to many	
Attributes:	No	**Yes**	

	Primary key field		Foreign key name
Origin table/feature class:	Georeference	:	Owner
Destination table/feature class:	Georeference	:	Property

1. Import the table TaxRecords2019 from the Tutorial 2-1 project folder (e.g., C:\EsriPress\FocusGDB\Projects\Tutorial 2-1) into the LandRecords geodatabase. Hint: Use the Geoprocessing pane to locate and run the Table to Geodatabase tool.

2. Set Input Table to TaxRecords2019 and Output Geodatabase to LandRecords, and then click Run.

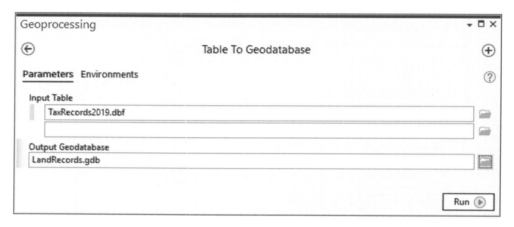

3. Start the creation of the relationship class by going to the Catalog pane and right-clicking the LandRecords geodatabase and clicking New > Relationship Class.

4. Using the information from the Relationship worksheet, enter the name of the origin table as Parcels, the destination table as TaxRecords2019, and the name of the output relationship class as Ownership.

5. Set the relationship type to Simple (peer to peer).

6. Type the labels that will describe the two relationships: Parcel is owned by for the forward path (as you traverse from the origin table) and Owner has ownership of for the backward path (as you traverse from the destination). Set Message Direction to None.

7. Set the relationship cardinality to M-N (many to many).

8. Check the box indicating that you want to add attributes for this relationship.

9. Use the two drop-down lists to set the origin and destination primary key fields to Georeference and GeoReferen, respectively. Then type the labels for the foreign key fields from the design form (see tutorial 1-1): Owner for the origin table and Property for the destination table. When your input matches the figure, click Run. Close the Geoprocessing pane when the process finishes.

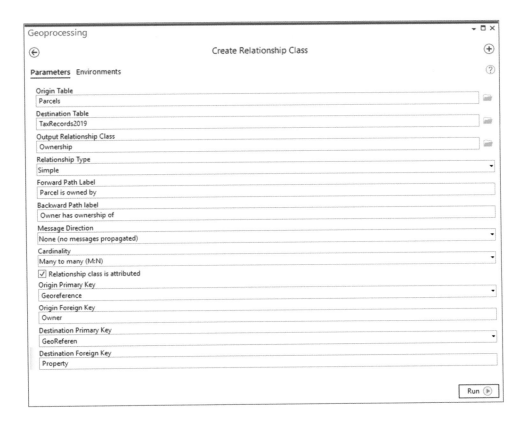

10. **Go to the Catalog pane and expand the PropertyData entry.**

The new relationship class has now been created in your geodatabase. The feature class Parcels is linked with the table TaxRecords2010 so that the information from the table can be used for analysis and symbolized through the parcels. As new tax rolls are released, more relationship classes can be built, allowing for comparisons of different tax years to be displayed in the Parcels feature class.

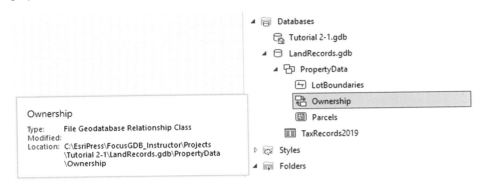

Test the subtypes

To get a better idea of the value of subtypes and the other data integrity rules, it would be beneficial to test them in ArcGIS Pro.

1. **If you have closed the project from earlier, open Tutorial 2-1 in ArcGIS Pro.**

2. **Expand the LandRecords geodatabase folder in the Catalog pane, and then find the PropertyData feature dataset. Right-click the feature dataset, and click Add to New > Map.**

   When a feature dataset is added to a map project, all the feature classes and tables in the feature dataset are added.

   The two feature classes LotBoundaries and Parcels now appear in the Contents pane under their aliases, Lot Boundaries and Property Ownership, respectively. Normally when a feature class is added to a map, it is symbolized with a single symbol. In the case of a feature class with a subtype, however, the data is automatically classified according to the subtype field. Notice that the Parcels layer shows the three categories defined as subtypes.

   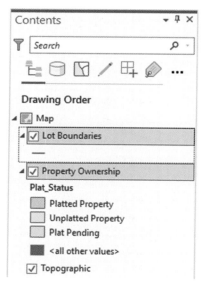

3. **Zoom to any place in the world you like. The coordinate system will be way off, but for this trial it'll be okay.**

4. **On the ribbon, click the Edit tab, and then click Create.**

   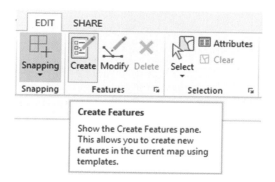

When the Create button is clicked, ArcGIS Pro creates a new editing pane named the Create Features pane. All the feature classes will appear in this pane along with their classifications. To edit, simply click the feature you want to work with, select an editing tool, and begin working in the map area.

5. **In the Create Features pane, click Platted Property, and then click the arrow on the right to open the attributes template.**

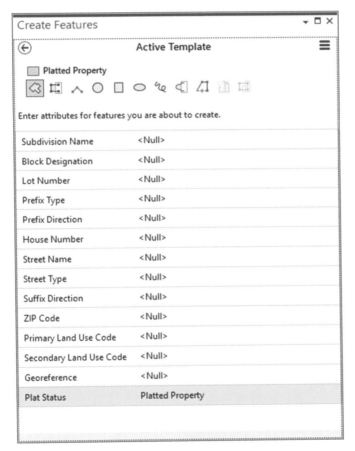

The attributes template will allow you to prepopulate the fields before features are drawn. Once populated, all new features will have these attributes. This process can be useful for attributes that will be common for a set of new features, such as subdivision name or street name. Also, any domains that were created for the feature class will be in effect when you prepopulate attributes.

6. **In the Active Template pane, fill in the following attributes:**
   - **Subdivision Name:** Bluffview Estates
   - **Street Name:** Meandering
   - **Street Type:** Way
   - **Primary Land Use Code:** Residential Property
   - **Secondary Land Use Code:** Single Family Detached

   Notice the domains that are in effect for Street Type, Primary Land Use Code, and Secondary Land Use Code and that the Plat Status field is already populated.

   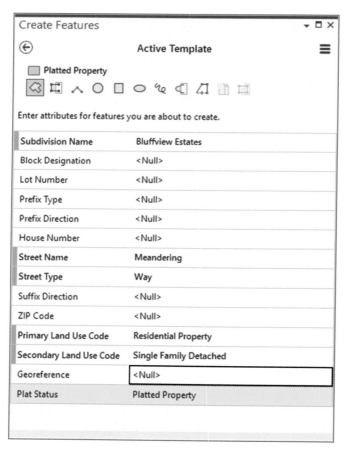

7. **Try drawing a few features in the map with the Platted Property subtype and then the Plat Pending subtype.**

   With Platted Property selected in the Create Features template, all new features drawn will automatically be put into the Platted Property subtype and all the unique data integrity rules you set up will be in effect. Without subtypes, you would have to go into the attribute table and change each entry as it is drawn.

The next tutorial will build more feature classes with data integrity rules. These features will be tested to give you an even better idea of the role of subtypes and domains.

8. **If you like, save the edits you have made. Save the map project, and then close ArcGIS Pro.**

Exercise 2-1

The tutorial showed how to create a geodatabase and all its components from the logical model. This design included a feature dataset, feature classes, domains, and subtypes.

In this exercise, you will repeat the process using the logical model for zoning data created in exercise 1-1.
- Start ArcGIS Pro, and locate the Tutorial 2-1 folder.
- Use the geodatabase worksheet Exercise 1-1 GDB Design.xlsx to create the geodatabase.
- Use the Tables worksheet to create the feature classes with the correct fields.
- Use the Domains worksheet to create and apply the domains.
- Use the Subtypes worksheet to create the subtypes.

WHAT TO TURN IN

If you are working in a classroom setting with an instructor, you may be required to submit the materials you created in tutorial 2-1.
- Screenshot image of the Catalog pane in ArcGIS Pro showing all the components of the completed geodatabases:
  - Tutorial 2-1
  - Exercise 2-1

Review

Any model of reality is only as good as the foundation on which it rests. This tutorial illustrated the development of a geodatabase using a systematic approach to design that documents the uses and relationships of geographic features of parcels in the City of Oleander.

The *enterprise geodatabase* is cross-platform compatible and can be operated using a variety of database management system (DBMS) formats, including Windows® SQL Server, PostgreSQL®, Oracle®, IBM® DB2®, and SAP HANA®. Typically used in larger organizations, an enterprise geodatabase can support many editors editing the geodatabase simultaneously and can efficiently manage these edits through versioning.

By using geodatabase design forms to design your geodatabase, you can clearly show specific representation and necessary relational behavior among geographic features. The use of the design forms provides a logical model of the geodatabase and focuses the developer's planning throughout the development process to ensure adequate representation of these features inside the geodatabase. As such, the development of a logical model is a central aspect of any geodatabase planning phase.

The project database in ArcGIS Pro was used to organize data in feature datasets, feature classes, and subtypes within the geodatabase. You can picture the hierarchical relationships of the geodatabase, feature datasets, and feature classes as a filing cabinet in which the geodatabase is the drawer of the filing cabinet, the feature dataset is the folder that resides in the drawer, and the feature classes are the individual pieces of paper within the file folder. You can assume that an organized relationship exists among these components in a filing cabinet if you ever want to find a document filed away. You can also assume that documents that are similar to each other, or that must support each other, are filed in the same folders. Remember, the rules of the geodatabase require that all feature classes supporting a topology or network must exist within the same feature dataset.

Data integrity and consistency issues within the LotBoundaries and Parcels feature classes of the geodatabase were addressed through the development of domains for several features. Proper use of domains in the beginning of design can save hundreds of hours of data cleanup from mistakes made during the data entry phase and countless hours of interpreting nonstandardized entries of data. Although domains are specifically used by feature classes, they are established at the geodatabase level to ensure that one central place will contain all domains and be available for all feature classes. Once a domain is created for a field within a feature class, the likelihood of making mistakes when entering data into this field is reduced dramatically, thus improving data integrity and consistency issues within the geodatabase.

During the initial planning phase of geodatabase development, you should examine the relationships among all feature classes and the features they contain with respect to their common behaviors and characteristics. Typically, you want to optimize (minimize) the number of feature classes contained in your geodatabase, while still capturing all the unique characteristics of each feature. This opportunity for optimizing your geodatabase is one of the advantages of using subtypes within a feature class. Creating subtypes for a feature class on the basis of certain defining characteristics of features within the feature class essentially allows you to build the flexibility of an additional feature class, without having to build and store one within your geodatabase. The City of Oleander used subtypes to classify the parcels it manages according to the status of the plat. The status of a parcel with respect to platting is critical to supporting the city's development activities. You could have created a separate feature class for each plat status. You would simply create a feature class for unplatted parcels, a separate feature class for platted parcels, and a third feature class for plat pending parcels. However, by recognizing that platted, plat pending, and unplatted characteristics of a parcel are simply a specific state of the evolution of the parcel in the land development process, you created a geodatabase with a subtype that would differentiate each parcel on the basis of its status. All three of these subtypes are still types of parcels. However, now they have been categorized in the geodatabase by a subtype. Regardless of what features you intend to store in a geodatabase, if the primary behaviors among features are consistent, it is good to consider a subtype instead of creating a separate feature class.

This tutorial also developed subtypes for the Parcels and LotBoundaries feature classes and illustrated another advantage of using subtypes when presenting your data to the user. When you displayed your Parcels feature class in a map within ArcGIS Pro, each feature class was automatically classified by these subtypes. Again, rather than adding three separate feature classes to display information on plat status within Oleander, the map used the subtypes from this one feature class to automatically classify features based on subtype. This is a more efficient way of working with and viewing the data. As you will see in later tutorials, the creation of subtypes gives you additional advantages when editing your data. Subtypes created within a feature class will allow each subtype to be a separate target feature during editing. This design gives the user the ability to use unique default values for each subtype and select unique values for each subtype from the supporting domain values. As you can see, a little thinking ahead regarding the use of subtypes and domains can help not only organize the behavior and characteristics of the features contained in your geodatabase, but also provide many advantages when displaying, editing, and populating the geodatabase with data.

STUDY QUESTIONS

1. List three advantages of using domains in your geodatabase. At what level within the geodatabase are these domains established, and why are they created at this level?
2. Why was it necessary to create parcels and lot boundaries feature classes? Could this characteristic have been represented by a subtype?
3. How does the establishment of default values within the table of a feature class benefit the geodatabase operation?
4. How can importing domains for a feature class benefit the geodatabase developer? Provide an example of suitable information that could be imported for a domain.

Other study topics

Search for these key phrases in ArcGIS Pro Help for further reading:
1. Create datasets in a geodatabase
2. Domains view - Geodatabases
3. Subtypes view
4. Fields, domains, and subtypes

Tutorial 2-2: Adding complex geodatabase components

The logical model included several techniques for simplifying data entry and managing data integrity. These features are included in the geodatabase when it is built in your project. If time is taken to design it well in the beginning, the long-term benefits will be great.

LEARNING OBJECTIVES
- Work with ArcGIS Pro
- Create a geodatabase
- Build a database schema

Introduction

Tutorial and exercise 2-1 worked with polygon feature classes. These feature classes have unique characteristics in their database design. The use of domains and subtypes helped build certain data integrity rules.

This tutorial will also focus on setting up feature classes with domains and subtypes but will work with linear and point data types. The process of setting up and using them is like the polygons used previously, so the tutorial instructions will move rather quickly. When everything is set up, however, a more extensive demonstration of the purpose of the data integrity rules will be presented.

Adding to the data structure

Scenario

You've completed designing and building the database for the parcel and zoning data for the City of Oleander. Now it's time to create the geodatabase structure for the sewer line designs that were also written earlier.

Data

Print out the design forms from tutorial 1-2. These forms portray the completed geodatabase logical model from which the sewer line schema can be created. If you have not completed tutorial 1-2, a finished set of forms is available in the Folders/Tutorial 2-2 folder.

Tools used

ArcGIS Pro:
- New file geodatabase
- New feature dataset

- New feature class
- Create a domain

Review the printed geodatabase design forms from tutorial 1-2. There are many components to build, and this tutorial will run through them rather quickly. If you encounter any problems, review the steps from tutorial 2-1.

Geodatabase design forms			ArcGIS Pro
Geodatabase name			Utility_Data
Feature dataset name			Wastewater
Feature classes:			
	Type	Feature class name	Alias
	LINE	SewerLines	Sewer lines
	LINE	Interceptors	Interceptors
	PNT	SewerFixtures	Sewer Fixtures
	PNT	InterceptorFix	Interceptor Fixtures

Create the data structure

The first part of the creation process will be to create the geodatabase and feature dataset. Check the design forms for the correct names.

1. **Start ArcGIS Pro, and open project Tutorial 2-2. Create a new file geodatabase named** Utility_Data. **Within this geodatabase, start the creation of a feature dataset named** Wastewater, **but don't set the spatial reference yet.**

   The last parameter to set is the spatial reference of the data to be stored here. It will be the same as the spatial reference for the Parcels feature class in the Tutorial 2-1 geodatabase, so it would be faster and more efficient to borrow a copy from that location.

2. **Click the Globe icon to select the coordinate system. Then using the drop-down arrow, click Import Coordinate System. Browse to the Parcels feature class in the Tutorial 2-1 geodatabase. Once the feature dataset is selected, click OK. Hint: If you set the FIPS 4202 coordinate system as a favorite before, expand the Favorites list and select it there.**

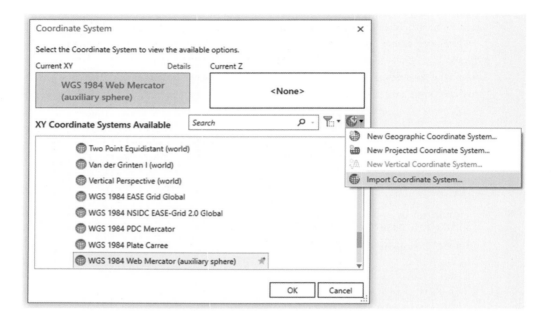

3. Finish the creation process by clicking Run.

   The Import process is a shortcut for setting the spatial reference. There are other situations where parameters can be imported from existing items, so always be careful to note which dialog screen is active. Only when the Spatial Reference dialog box is open will the Import process import a spatial reference.

4. Go to the Wastewater feature dataset and start the creation process for the SewerLines feature class, being very careful to set the feature type correctly. You may decide whether you want to enter the fields and their parameters in this feature class creation pane or wait until the next step and enter them through the spreadsheet forms. Remember that you will design and apply the domains and subtypes later. When it is complete, click Finish, and close the Geoprocessing pane.

Feature class or table name	Field name	Alias	Field type	Nulls (Y/N)	Domain name or subtype field (S) or (D)	Default value	Length
SewerLines	Pipe_Size	Pipe Size	SI	N	(D1) Sewer_Pipe_Size		
	Material	Pipe Material	SI	N	(S1) Sewer_Line_Material		
	Year_Built	Year Built	LI	Y			
	Description	Owner	Text	N			

5. (Skip this step if you entered the fields in the feature class creation pane.) Right-click the new feature class, and click Design > Fields. In the design dialog box, enter all the fields from the design form. Pay close attention to the Allow NULL setting and any default values. When all fields are entered, click Save on the Field tab, in the Changes group.

The field Pipe Size has a domain. Look at the Domains worksheet to confirm the entries. The traditional process for domains is to create the table, open the design interface of the geodatabase and create the domain, and then go back to the feature class field design pane and apply the new domain. There is, however, a faster way to create domains through the field design pane that you already have open.

Domains worksheet

	Domain name	Description	Field type	Domain type	Coded values / Range			
					Code (Min)	Value (Max)		
D1	Sewer_Pipe_Size	Sewer Pipe Size	SI	Coded Values	6	6"		
					8	8"		
					10	10"		
					12	12"		

6. Open the field design pane for the SewerLines feature class (if not already open from step 5). Double-click the Domain column next to the PipeSize field. In the drop-down box, click Add New Domain.

7. Fill in the parameters for the domain, following the design sheet. When you're finished, save the domain (go to the Domains tab, and click Save), and close the Domain design pane.

8. When you return to the Fields design pane, you will notice that the new domain is now available in the drop-down selection for the field Pipe_Size. Set the domain to Sewer_Pipe_Size.

9. **On the Fields tab, in the Changes group, click Save, and close the pane.**

In the next feature class, the fields are the same as in the SewerLines feature class. The feature class design pane will allow you to use an existing feature class as a template that will make filling out the fields dialog box simple.

Import field definitions

The next feature class on the design form is interceptors. But because you will want to use an existing feature class as a template for creating the new one, you will use the Create Feature Class geoprocessing tool, which has that option.

Feature class or table name	Field name	Alias	Field type	Nulls (Y/N)	Domain name or subtype field (S) or (D)	Default value	Length
Interceptors	Pipe_Size	Pipe Size	SI	N			
	Material	Pipe Material	Text	N			
	Year_Built	Year Built	LI	Y			
	Description	Owner	Text	N			

1. **Open the Geoprocessing pane. Then find and open the Create Feature Class tool.**

2. **Drag the feature dataset Wastewater into the dialog box and into the Feature Class Location box. Enter the feature class name, and set the geometry type as** Polyline **(which is the selection for Line). Then set the template feature class to** SewerLines. **Note that you can drag the feature class from the Catalog pane to the Geoprocessing pane to populate the entry. You do not need to set a coordinate system because it will automatically be inherited from the feature dataset. Click Run to create the feature class, and then close the Geoprocessing pane.**

3. **Open the Fields design pane for the new feature class Interceptors.**

   You will see that all the fields have been copied from the SewerLines feature class, and even the Sewer_Pipe_Size domain was set. However, one of the field types is different from the imported data structure. This field can be changed while you are working in the design pane. Note that changing a field type after data has been entered in a feature class may result in loss of the data, but since this one is empty, there are no concerns.

4. **Set the data type for the Material field to Text as shown on your design form. Save the changes, and close the Fields design pane.**

Continue the creation process

Now move on to the feature class for the sewer fixtures. You may decide which method you want to use for the feature class creation.

Feature class or table name	Field name	Alias	Field type	Nulls (Y/N)	Domain name or subtype field (S) or (D)	Default value	Length
SewerFixtures	Fix_Type	Fixture Type	SI	N	(S2) Sewer_Fix_Type		
	Flowline	Flowline Elevation	Float	Y			
	Rim_Elev	Rim Elevation	Float	Y			
	Depth	Depth from Surface	Float	Y			
	Year_Built	Year Built	LI	Y		2019	
	Description	Owner	Text	N		Oleander	
	Material	Pipe Material	Text	N			
	Year_Built	Year Built	LI	Y			
	Description	Owner	Text	N			

1. **In the Wastewater feature dataset, create the** SewerFixtures **feature class. Add the fields and their parameters from the Design worksheet. In this instance, you can go ahead and set the default values for Year_Built and Description. The domains will be set up later. When the field settings are correct, save the field changes and close all the dialog panes.**

   The final feature class to create is for the interceptor fixtures. Notice that it has four of the same fields as the sewer fixtures.

Tables worksheet							
Feature class or table name	Field name	Alias	Field type	Nulls (Y/N)	Domain name or subtype field (S) or (D)	Default value	Length
InterceptorFix	Flowline	Flowline Elevation	Float	Y			
	Rim_Elev	Rim Elevation	Float	Y			
	Depth	Depth from Surface	Float	Y			
	Year_Built	Year Built	LI	Y		2019	

2. **Once again, go to the Wastewater feature dataset, and create a new feature class for** InterceptorFix. **While you are creating it, use the template option to transfer all the fields from the feature class SewerFixtures.**

3. **Two of the fields are not necessary. Remove them by opening the field design pane, clicking the gray cell on the left of their name, and pressing Delete. The field design pane will show them with a line struck through them until you save the changes. After you save it, close the dialog box.**

Field Name	Alias	Data Type	Allow NULL	Domain	Default	Length
OBJECTID	OBJECTID	Object ID	☐			
Shape	Shape	Geometry	☑			
~~Fix_Type~~	~~Fixture Type~~	~~Short~~	☐			
Flowline	Flowline Elevation	Float	☑			
Rim_Elev	Rim Elevation	Float	☑			
Depth	Depth from surface	Float	☑			
Year_Built	Year Built	Long	☑		2019	
~~Description~~	~~Owner~~	~~Text~~	☐		~~Oleander~~	~~255~~

Now go back and check your work. Did you create all the feature classes as the correct feature types? Did you get all the Allow NULL settings correct? Did you spell all the field names correctly? If not, now is the time to make the changes so that you will not hit any interference later when you add data. Also, check field alias names and required default value settings. If any of these settings are incorrect, you can change them now.

Create subtypes

The last of the data integrity rules involves the subtypes. Review the Subtypes worksheet.

	Subtype name	Code	Description	Field	Domain name	Default value
S1	Sewer_Line_Material	1	PVC	Pipe_Size	Sewer_Pipe_Size	8
				Description		Oleander
				Year_Built		2019
		2	HDPE	Pipe_Size	Sewer_Pipe_Size	10
				Description		Oleander
				Year_Built		2019
		3	DI	Pipe_Size	Sewer_Pipe_Size	12
				Description		Oleander
				Year_Built		2019
		4	Conc			
		5	Clay			
S2	Sewer_Fix_Type	1	Manhole			
		2	Cleanout			

The second subtype is less complex, so for learning purposes, you should create that one first. Then a more detailed description will follow to create the more complex subtype.

1. Right-click the SewerFixtures feature class, and click Design > Subtypes.

2. In the Subtypes pane, right-click Fix_Type, and click Set as subtype field. This step not only sets the subtype field but also opens the Manage Subtypes pane so that you can enter the subtype values. Note that the subtype field will now be bold and have an asterisk next to the name.

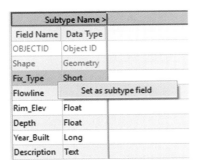

3. Follow the design sheet as a guide, and enter the codes and descriptions for the subtype. When the values are entered, click OK and Save to complete the process.

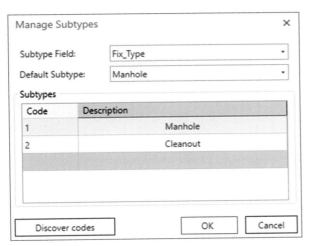

The Manage Subtypes pane has an interesting option: Discover codes. If there was already data in this feature class, this option would find every unique value in the subtype field and make a code for it. You would just add the description and be finished. This option is useful when you are importing existing data and want to retroactively add a subtype.

Now to tackle the tougher subtype. This one has codes and descriptions like the other subtype but also includes a domain and default setting for each subtype code.

4. Right-click the SewerLines feature class, and open the Subtypes design pane. Then right-click the Material field, and set it as the subtype field.

5. Use the design sheets for reference, and enter all the codes and descriptions. Click OK to save the codes before moving on. Don't worry about the defaults and domains; these options are set after the codes are saved.

6. In the Subtypes pane, enter the defaults for Pipe_Size, Year_Built, and Description. Click Save, and close the Subtypes pane.

Subtype Name >		P.V.C.		HDPE		DI		Conc		Clay	
Field Name	Data Type	Domain	Default Value	Domain	Default Value	Domain	Default Value	Domain	Default Value	Domain	Default Value
OBJECTID	Object ID										
Shape	Geometry										
Shape_Length	Double										
Pipe_Size	Short	Sewer_Pipe_Size	8"	Sewer_Pipe_Size	10"	Sewer_Pipe_Size	12"	Sewer_Pipe_Size		Sewer_Pipe_Size	
*Material	Short		1		2		3		4		5
Year_Built	Long		2019		2019		2019				
Description	Text		Oleander		Oleander		Oleander				

This work wraps up the database creation process. All the components of the logical model have been entered. On complex datasets, it's a good idea to run back through the logical model and check the entries against the feature classes to make sure nothing was missed. If you were methodical in the process, everything should be okay.

7. Save the ArcGIS Pro project.

Test the rules

As in tutorial 2-1, it is advisable to test the features in ArcGIS Pro before proceeding. This quick pilot study will determine whether any of the rules aren't working and will give you the chance to go back into the data structure and correct them if necessary. Once data is loaded into the feature classes, it would be much riskier to try to fix problems, especially if it involved renaming a field or changing a data type.

1. If you closed ArcGIS Pro after the first part of this tutorial, open it now using the Utility_Data project. Add the feature dataset Wastewater from the Utility Data geodatabase to the project, making a new map. All the feature classes in the Wastewater feature dataset should be added to the Contents pane.

Chapter 2 | Creating a geodatabase 73

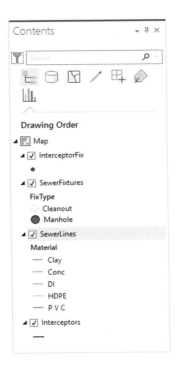

Notice that the layers SewerFixtures and SewerLines have already been classified by their subtype values. ArcGIS Pro does this classification automatically, although if it is not the desired classification, you can change it manually.

2. Click the Edit tab, and in the Features group, click Create. The Create Features pane will display all the editable features, grouped by feature class and subtype. Select PVC from the SewerLines listing.

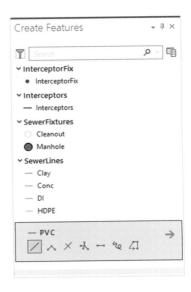

Notice that the list of selections under SewerFixtures and SewerLines comes straight from the subtypes you defined earlier.

3. **Use the default construction tool in the map area to draw a sample line, and complete it by double-clicking the last point. As before, you can try prepopulating some of the attributes before drawing features. Try adding several of the different features, and then save the edits when you are finished.**

Notice in the attributes dialog box that all the fields have already been filled in! These entries are the default values that you defined in the layer properties. For the SewerLine type of PVC, the default size is 8", the year built is 2019, and the description regarding ownership of the line is set to Oleander.

YOUR TURN

Draw each type of sewer line, and notice the attribute values that are assigned to each one. Remember that the defaults apply only to new items drawn. Since concrete and clay pipes are rarely installed as new, there were no defaults set for these types. They needed to appear in the database, however, to accommodate existing pipes made of these materials. If they were not in this list, they would test as invalid when the database was checked.

4. **When you have finished working with the various wastewater features, save your edits and close ArcGIS Pro, making sure to save the project.**

This was a quick test of the data structure. If any errors were found, you could go back and correct them, but make sure to save any edits first before making any corrections. Remember that once data is added to these feature classes, it will be more difficult to change the data structure after the fact.

Exercise 2-2

The tutorial showed how to create a geodatabase and all its components from the logical model. These components included a feature dataset, feature classes, domains, and subtypes.

In this exercise, you will repeat the process using the logical model for a storm drain system created in exercise 1-2.

- Start ArcGIS Pro, and open the Tutorial 2-2 project.
- Use the geodatabase worksheet to create the geodatabase.
- Use the tables worksheet to create the feature classes using the correct fields.
- Use the Domains worksheet to create and apply the domains.

- Use the Subtypes worksheet to create the subtypes.
- Test the results by adding the feature classes to a new map.

WHAT TO TURN IN

If you are working in a classroom setting with an instructor, you may be required to submit the materials you created in tutorial 2-2.
- Screenshot image of ArcGIS Pro showing all the components of the completed geodatabase:
  - Tutorial 2-2
  - Exercise 2-2

Review

This exercise built on the concepts of creating unique attribute domains and subtypes for linear features. The City of Oleander is managing its sewer network in a geodatabase. As you can see, the same process that was used for setting up domains and subtypes for parcels was used for the sewer system. By using the standardized planning process and filling out the forms ahead of time, you were able to simplify creation of the geodatabase for the sewer system.

ArcGIS Pro provides all the necessary tools to easily create your geodatabase; however, careful thought was necessary to help establish the behaviors and relationships for these features. As you could see, separate feature classes could have been created for the different materials of pipes and fixtures. However, regardless of the material of pipe or fixture, each one has the same basic function and could be managed more efficiently through the creation of subtypes based on material. Using subtypes to define categories of features is common in the industry to distinguish behavior, attributes, and access properties of features. A domain was created for the size of the sewer pipes to help with the standardization of these values to enforce data integrity.

As you have worked through the tutorials, you have probably noticed that there has always been a step to test each component that you built into your geodatabase to ensure that it represents the features as the needs of the project dictate. This QC exercise is a critical step that should be conducted as often as possible throughout the development project. Quality control of your database development project should not be skipped because of time constraints or for any other reason. There is nothing worse than completing the construction of a project and realizing that it does not adequately meet the needs of your client because of a logical oversight or because an error was made early in the development process. Typically, these situations can surface in two different ways that you can help control with thorough planning: either the features, attributes, behaviors, access restrictions, and so on, built into the model do not adequately support the workflow process they were designed to support or errors were made in the execution of the plan. It's simple. When a failure occurs, you either have a bad model or poor execution. (Of course, this scenario assumes that the workflow process itself is designed adequately to achieve the necessary results. Most likely you will have no control of the process that you are hired to support.)

To help you succeed with your database design project, do the following:

1. Meet with, and document your interviews with, project sponsors to discern the necessary requirements for the project.
2. Have the project sponsor review and approve the project requirements document that you compiled.
3. Use the geodatabase design forms to carefully think through an appropriate approach that reflects the approved results from your requirements session.
4. Review your design with the project sponsor and get his/her approval of your proposed design.
5. Carefully execute your plan for developing your geodatabase.
6. Conduct testing to ensure that your geodatabase meets your requirements.
7. Implement your geodatabase.

Many of the fastest ways to fail in a GIS database development project can be avoided simply by carefully checking your work, having a colleague check your work, and then rechecking your own work one final time. One of the advantages to using the design forms during the planning phase is that you create a documented paper trail of your logic that you can refer to if a failure occurs and you need to troubleshoot your design. The geodatabase provides many of the tools you will need to help build an efficient, flexible, accurate, and consistent model of reality. In the last two chapters, you have learned how to organize your geographic features in a geodatabase and how to incorporate attribute domains and feature subtypes into the geodatabase. These tools, as well as the many others that you will be learning in upcoming tutorials, will help you become successful, but remember that the most critical component of any GIS is *the capable mind of a thinking operator*.

STUDY QUESTIONS

1. Explain the rationale for creating subtypes for the sewer lines. What other approach could have been used, and how? What situations prohibit the use of subtypes? What advantages do subtypes offer during data editing?
2. What are the primary advantages of creating domains for the sewer pipes? How will this step save you time later?
3. Explain the relationship between feature datasets and feature classes. What are the advantages and disadvantages of grouping your feature classes into several feature datasets?

4. List three mistakes that can be made when creating a feature class that would cause you to have to re-create the entire feature class. How can these mistakes be avoided?
5. What is the most critical component of any geodatabase?

Other study topics

Search for these key phrases in ArcGIS Pro Help for further reading:
1. Define feature class properties
2. Create feature class
3. Subtype group layers
4. Geoprocessing considerations for subtypes

Chapter 3

Populating and sharing a geodatabase

Your data may come from other sources in a perfectly formatted geodatabase, or you may be tasked with designing and creating the perfect geodatabase yourself (as you've been doing in the first few tutorials). Either way, you will reach a point where you must get new data into your existing geodatabase in the most efficient way possible, taking full advantage of the special data integrity rules that you have set up. When all of this is finished, you may also be tasked with sharing this data so that others can reap the benefits of your hard work. This chapter will help you explore some of those capabilities.

Tutorial 3-1: Loading data into a geodatabase

Lots of data sources exist, and any one of them can be brought into a geodatabase. This first look at importing data will deal with bringing data from a polygon shapefile into a geodatabase.

LEARNING OBJECTIVES
- Import data
- Work with shapefile data
- Create a load procedure
- Load data into a subtype

Introduction

There are lots of ways to put data into a geodatabase format. The simplest is to find the data in ArcCatalog, right-click it, and click Export > To Geodatabase or use the Geoprocessing tool in ArcGIS Pro called Shapefile to Geodatabase. With either method, however, you have no control over the data structure of the output file. The file will have the same data structure as the input data, and it won't have the built-in data integrity rules that you designed into the data in the last few tutorials in chapters 1 and 2.

A better process is to load data carefully into the new geodatabase structure, considering the different feature classes and their subtypes. This manual loading is done by putting the source data into your ArcGIS Pro project, and then carefully selecting data and placing it into the subtypes that are part of your geodatabase design. It's the difference between dumping a wheelbarrow full of data into your computer versus making your own containers and carefully selecting what goes into each one. It's important to know that any mistakes in loading the data cannot be undone, so all the imported data would have to be deleted and the process started again. Because of this possibility, you should be careful to set each parameter correctly for each load procedure.

The loading process can also be tedious, requiring that you repeat the process many times to separate the data into a new framework. Imagine one big dataset that contains all the water, sewer, and storm drain utilities. Your new design might split the data into four or five feature classes, each with several subtypes. You would need to run the loading process for each feature class/subtype combination, which could mean that you run the process 20 or 30 times before all the data is incorporated into the new data structure. But once it is completed, the new data integrity rules and other functions built into the geodatabase structure will make editing and updating much more streamlined.

Establishing a load procedure

Scenario

Having gone through the design process for the parcel data and come up with a good design, you can now populate the new geodatabase and start using the new format. After digging through the archives, city staff discovered some shapefiles containing the data you need. You will use ArcGIS Pro to select and place this data into the new data structure.

Data

The project Tutorial 3-1 already has an empty geodatabase that matches the data schema from tutorial 2-1. This geodatabase is where you will store the data. The source shapefiles in the Tutorial 3-1 folder are called ParcelLineSource.shp and ParcelSource.shp.

The ParcelSource.shp file contains a field named PlatStatus, which will be used to match the data to the subtypes:

1 = Platted Property
2 = Unplatted Property
3 = Plat Pending

The other fields for ParcelSource.shp, which will be used to match the source fields to the new geodatabase, are as follows:

Prop_Des_1	subdivision or abstract name
Prop_Des-2	lot and block designation
Acreage	area of parcel measured in acres
DU	number of dwelling units on parcel
PlatStatus	status of the platting process
PIDN	tax office identifier
Prefix	street prefix (N,S,E,W)
StName	street name
Suffix	street suffix

SufDir   street suffix direction (N,S,E,W)
LotNo   lot number
BlkNo   block number
PID1   subdivision code
PID2   lot or tract number
PID3   tax transaction code
PID4   tax transaction code
Prop_Add   situs address
Addno   address number
EKEY   tax account number
GeoReferen   tax office georeference key
YearBuilt   year of construction
Marker   internal use
TADMap   tax office map number
PoliceDist   police district number
RandomNum   random parcel selection
HouseValue   property appraised value
CityOwned   city-owned property
Shape_Leng   perimeter of parcel (provided by ArcGIS)
Shape_Area   area of parcel (provided by ArcGIS)

Field Name	Data Type
FID	Object ID
Shape	Geometry
Prop_Des_1	Text
Prop_Des_2	Text
Acreage	Float
DU	Short Integer
PlatStatus	Short Integer
PIDN	Text
Prefix	Text
StName	Text
Suffix	Text
SufDir	Text
Lotno	Text
BlkNo	Text
PID1	Text
PID2	Text
PID3	Text
PID4	Text
Prop_Add	Text
Addno	Long Integer
EKEY	Long Integer
GeoReferen	Text
YearBuilt	Short Integer
Marker	Text
TADMap	Text
PoliceDist	Text
RandomNum	Long Integer
HouseValue	Double
CityOwned	Text
Shape_Leng	Double
Shape_Area	Double

The source file ParcelLineSource contains linear features that represent the outlines of the property ownership parcels. This data will be put into the Lot Boundaries feature class. If you recall from the design forms, the Lot Boundaries feature class has a text field to store a line code. The source data has a similar field as an integer to represent the same categories.

Tools used

ArcGIS Pro:
- Select Layer By Attribute
- Append tool

The process begins with looking at the existing data and deciding which parts will go where in the new data structure. The old data may have more fields than needed, but it may also be missing data that was included in your design. You may also find that data that was stored as an integer code will not go into a text field with a better description, such as the Lot Boundaries codes. Once a field map is created, the process will become clear.

Create a field map

The field map will be a list of all the fields from your new data structure with their counterparts listed from the source data.

1. Print a copy of the completed tables form from the geodatabase design forms you made for parcels in tutorial 1-1. These forms can be your own design sheets or the ones provided with tutorial 2-1.
   On the right of each field, write the name of the source field for that data using the field list for the ParcelSource.shp file and trying to match the field descriptions. Write your own first, and then check it against the figure.

Output Field Name	Source Field Name from Shapefile
Sub_Name	Prop_Des_1
Blk	BlkNo
Lot_No	LotNo
Pre_Type	<None>
Pre_Dir	Prefix
House_Num	Addno
Street_Name	StName
Street_Type	Suffix
Suffix_Dir	SufDir
ZIP_Code	<None>
Primary_Use	<None>
Secondary_Code	<None>
Georeference	GeoReferen
Plat_Status	PlatStatus

That solves the question of which fields match up as far as data loading, but there are also the subtypes to consider. In the source data, Plat Status is shown by an integer value of 1, 2, or 3. As you load the data, you'll use a query to select each of these codes and load it into the subtype you set up in tutorial 2-1.

Start the loading process

1. Start ArcGIS Pro, and open the project Tutorial 3-1.

2. In the Catalog pane, expand the Folders > Tutorial 3-1 folder to see the items contained there.

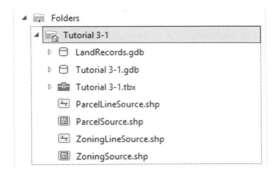

The project contains several shapefiles, two of which contain the parcel data that you will work with. It would be a good idea to place these shapefiles in the map and separate them into a group layer to distinguish them from the destination data.

3. In the Contents pane, right-click the map name (in this case, Oleander Property), and click New Group Layer. Rename the group layer *Source Data Group Layer*.

Because group layers can often have the same characteristics and behaviors as a data layer, it's always a good idea to include the words *Group Layer* in your group layer names. This naming convention will help remind you that there may be several layers involved with this single Contents pane entry.

4. Drag the ParcelLineSource and ParcelSource files from the Catalog pane into the new group layer, and make sure they are visible.

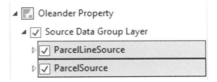

The process to load the parcel polygon data will involve using the Append tool to load that data into the Parcels feature class. Recall that the Plat Status codes are as follows:
  1 = Platted Property
  2 = Unplatted Property
  3 = Plat Pending

You will need to find and run the Append tool to transfer features to the Parcels feature class. The tool will allow you to select a subset of the data and load it directly into a subtype. For instance, the first time you run the Append tool, you will select the platted property from the shapefile and place it in the platted property subtype in the destination feature class. This selection process will be done once for each of the three categories.

5. **Open the Geoprocessing pane, and then find the Append (Data Management) tool and open it.**

6. **Set Input Datasets to Source Data Group Layer\ParcelSource, and set Target Dataset to Property Ownership.**

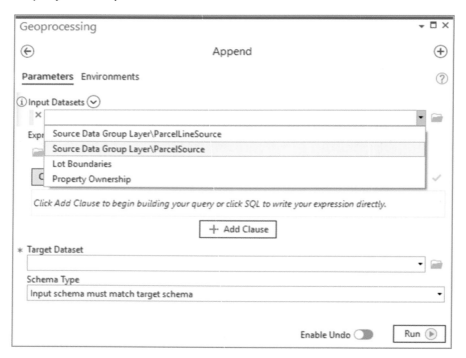

7. Change the schema type to "Use the Field Map to reconcile schema differences." Note that a new Field Map dialog box appears.

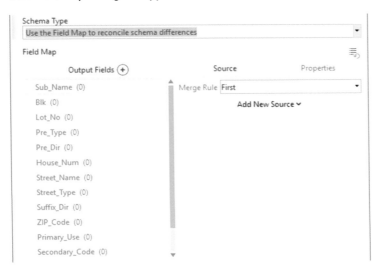

Now is when the field mapping comes into play. The box lists the output fields and asks you to identify the source fields. You'll use the diagram you made in the first step to fill in the target field matrix.

8. In the Output Fields column, click Sub_Name. In the Source column, click Add New Source, and then scroll down and check the field Prop_Des_1. Click Add Selected to make the field assignment.

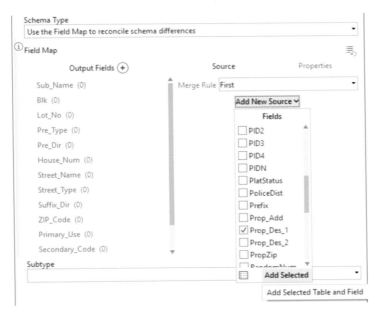

YOUR TURN

Continue down the list from the field map, selecting each output field and matching it with a source field. Set each source field to match the field map you created at the beginning of the tutorial. The output fields will turn black when they are correctly matched. Some of the fields do not have a match, so they will remain red.

Note the Enable Undo option at the bottom of the pane. Enabling this function would allow you to undo an incorrect Append action. For instance, if you forgot to do the selection before running the tool or incorrectly matched a field, you could undo and try again.

9. The final step is to set the subtype into which this data will load. Click the Subtype drop-down arrow, and select Platted Property.

10. When everything is set correctly, click Run. Do not close the Geoprocessing pane when this process is complete.

The load won't take long. If you want to see the results, you can open the attribute table of Property Ownership and view the populated fields.

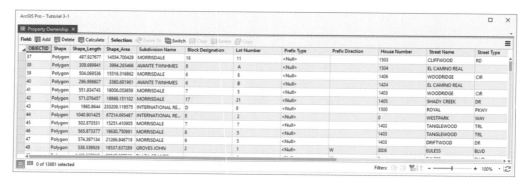

You've created the load procedure that has loaded the first subtype. There are two more to load: the unplatted property (Plat_Status = 2) and the plat-pending property (Plat_Status = 3).

YOUR TURN

Setting up for the next data loading is easy. You will need to modify the query to find only the records where Plat_Status = 2, and change the destination subtype to Unplatted Property. The field mapping will stay the same. Make these changes, and run the Append tool again.

Finally, change the query to find the Plat-Pending property and place it in the correct subtype. When all the data is loaded, you can close the Geoprocessing pane.

Check the results

1. In the Contents pane, turn off the Source Data Group so that only the Parcels feature class is visible.

2. Zoom and pan around the data, and notice that it is color-coded by the subtype field Plat Status.

3. Open the attribute table for Property Ownership, and notice that all the data was loaded into the correct fields.

Load the linear data

The process of loading the polygons that represent parcels is completed. Next, you will load the data that represents the edge of the parcels. Tutorial 1-1 explained that because the edges may need to be symbolized individually, they must be in their own feature class. In chapters 5, 6, and 7, you will learn how to manage the association between these features using other data integrity rules.

Recall from the data description at the start of this tutorial that there are three integer values for the Line Code field in the source data:

1 = Lot Boundary
2 = Right-of-Way Boundary
9 = Split Lot Line

These values will be matched to the domain values for the new feature class Lot Boundaries, and the values will be changed from an integer to text.

LOT = Lot Line
ROW = Edge of Right-of-Way
SPLIT = Split Lot Line

The Append tool will be run once for each input value, and then the field value will need to be recalculated from the integer code to the text code to match the domain.

1. Open the Geoprocessing pane and find the Append tool. Note that it is also in the Tools pallet under the Analysis tab.

2. Set Input Datasets to Source Data Group Layer\ParcelLineSource and Target Dataset to LotBoundaries.

3. **Add the query expression** Where Linecode is equal to 1 **to load only the data for lot lines.**

4. Change the schema type to "Use the Field Map to reconcile schema differences." There is only one field to map: Line_Code in the output fields to Linecode in the source fields.

5. Note that there is no subtype option to set for this feature class. Click Run.

The value that was loaded is 1, but the value in the new feature class should be ROW to match the domain. You can fix this discrepancy using the Field Calculator tool.

6. In the Geoprocessing pane, find and run the Calculate Field tool.

7. Set the input table to Lot Boundaries, the field name to Line Code, and the output value to "LOT". Click Run.

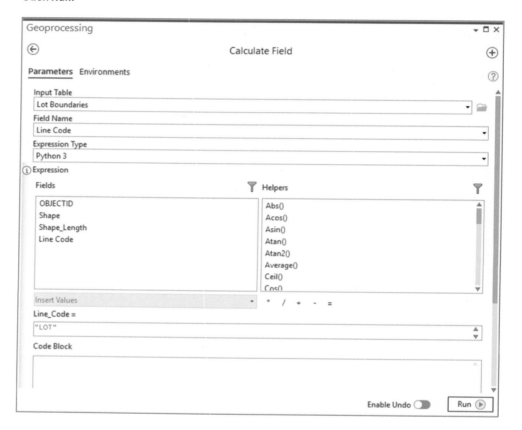

YOUR TURN

Repeat the append process for the values of 2 and 9. Remember that each value will require a unique expression when the data is loaded. Next, select the features with a Line Code value of 2 and recalculate their value to "ROW". Finally, select the features with a Line Code of 9 and recalculate their value to "SPLIT". Note that you will need to run the Select Layer By Attribute tool for each set of data.

The Lot Boundaries layer didn't automatically symbolize by the Line Code field. Set a new Symbology class for Unique Values using the Line_Code field. Make code LOT a thin line, code ROW a thicker line, and code SPLIT a dashed line.

All the data has been successfully loaded into the geodatabase. This is where the data will be stored and maintained for many years to come. If you like, zoom in and pan around the data and notice how the features are set up. The geodatabase has many data integrity rules built into it but retains the flexibility to be modified to accommodate future expansion and analysis.

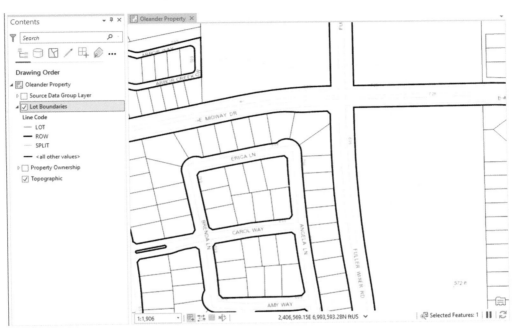

8. Save and close the ArcGIS Pro project.

Exercise 3-1

The tutorial showed how to use the Select Layer By Attribute and Append tools to selectively add data to an existing geodatabase/feature class structure.

In this exercise, you will repeat the process and load data into the zoning geodatabase you created in exercise 2-1. This data will include both the polygons and lines that represent the zoning districts and their boundaries.

- Start ArcGIS Pro, open project Tutorial 3-1, and go to the ZoningData geodatabase that you created in exercise 2-1.
- Use the Geoprocessing pane to find the necessary tools, and load the zoning data into this geodatabase. The two files containing the source data are in the Tutorial 3-1 folder. The file ZoningSource.shp contains the polygon information, and the file ZoningLineSource.shp contains the zoning boundary information. Investigate their data structure if necessary

to determine the best procedure for importing the data. Remember to import into the subtypes you created.
- When you're finished, create a new map layout in ArcGIS Pro and test the data integrity rules.

WHAT TO TURN IN

If you are working in a classroom setting with an instructor, you may be required to submit a screenshot of the geodatabase you loaded in tutorial 3-1.
- Screenshot image of the geodatabase previews from:
  - Tutorial 3-1 (geography and table)
  - Exercise 3-1 (geography and table)

Review

Data creation is one of the costliest aspects of GIS implementation that organizations face. Consequently, the ability to import existing data from other digital formats into your geodatabase has the potential for significant cost savings if it's done correctly. Tools contained in ArcGIS Pro and ArcCatalog have proven to be successful in importing existing digital data from other formats into the geodatabase. One of the key reasons that these tools are successful is because of the flexibility they provide to customize the import of geospatial data on the basis of user needs and the geospatial data structure and format. As illustrated in the last two chapters, developing a good plan to import your data is crucial to the success of building your geodatabase.

ArcGIS Desktop (both ArcGIS Pro and ArcCatalog) offers a variety of import capabilities for existing shapefiles. One way to import shapefile data into your geodatabase is to use the Feature Class To Geodatabase tool to create a feature class from the existing shapefile features and attributes. Although this option is the most expedient in the short term, it may cost you later when you attempt to enhance the functionality, standardization, and security of your geodatabase according to established user requirements. Another way to import data is to use ArcGIS Pro and the Append tool to import your data into the structured design of the existing geodatabase you created. To accomplish this process in the exercise, you created a load procedure that instructed the Append tool to "match" or "map" the import shapefile data schema to existing fields and subtypes in the geodatabase. As you have already devoted quite a bit of time to organizing your data according to feature types, relationships, and desired functionality, you will save time by transferring import data directly into the geodatabase according to the rules you established in the load procedure.

By using the Append tool in ArcGIS Pro and its field mapping capabilities to assign each feature and attribute from the import source data to its appropriate feature type and subtype in the output feature class, you were able to use your established geodatabase design to import data into the geodatabase. Although this method initially requires more time than using an automated means of importing your data into the geodatabase, it pays for itself in the long run. As in the initial planning activities during the geodatabase design phase in chapter 1, the Append option of importing data was enhanced by a "load procedure" to determine where features and attributes of the import tables will reside in the geodatabase. This process may involve multiple iterations until every feature and subtype is appropriately "matched" to the existing geodatabase structure.

STUDY QUESTIONS

1. Name two options for transferring data from an existing shapefile into a geodatabase. Explain the advantages and disadvantages of each option.
2. Considering what you already know about feature subtypes, why is it important to have the ability to import features directly into a subtype of the existing geodatabase?
3. Why was it important to select features from the ParcelSource file and selectively place them into the subtypes? What would happen if you just ran the Append tool without specifying the subtype?

Other study topics

Search for these key phrases in ArcGIS Pro Help for further reading:
1. Append Data
2. Append Data expressions

Tutorial 3-2: Populating geodatabase subtypes

Loading data from various sources is achieved in the same manner, regardless of what the data represents. The feature types and fields may be different, but they are still loaded into a geodatabase using the Geoprocessing tools and carefully crafted queries.

LEARNING OBJECTIVES

- Work with shapefile data
- Create a load procedure
- Load data into subtypes

Introduction

Now you've seen how the load process works. Although it may be tedious, it is the best way to precisely control how your data is loaded into a geodatabase. The linear and point features here will load the same way as the polygon and linear features in the last tutorial. Be diligent in setting the subtypes and queries to make sure everything loads correctly.

Loading the data into multiple feature classes

Scenario

You completed the geodatabase design for the sewer system in chapter 1, created the geodatabase in chapter 2, and now will be populating it with previously created sewer line data that the city had from a previous mapping endeavor. As before, you'll use various geoprocessing tools to carefully move the data from the shapefile to the geodatabase. This time, however, the data will be loaded into several feature classes from the same data source, and subtypes will be used to segregate the data even further. Be careful to check all the parameters and queries you'll be using in the Select Layer By Attribute and Append tools to ensure success. Any misloaded data will be cause to start all over.

Data

You already have the empty containers ready—the Utility_Data geodatabase created in tutorial 2-2. The existing shapefiles with the sewer system data are named SewerLineSource.shp and SewerNodeSource.shp. They are in the Tutorial 3-2 folder.

SewerLineSource.shp contains the linear data for the pipes. It will be loaded into subtypes in the new feature class using the material type. It will also be split between two feature classes, one for city-owned pipe and one for interceptors that are owned by a regional treatment facility. The following field descriptions will be used to match the existing data to the new field structure:

   PSize: Pipe size
   Material: Pipe material (PVC, HDPE, DI, Conc, Clay)
   Year_Const: Year of construction
   Comments: Notation of pipe's function
      Oleander = standard collection pipe
      Interceptor = large-size interceptor pipe

SewerNodeSource.shp contains the data for the manholes and cleanouts, as referenced in the designs. Once again, be aware of the subtypes, and note that the data will be split into two feature classes, in the same way that the pipes were.

   SYMB: Code for manholes or cleanouts
      11 = manhole
      15 = cleanout
   Depth: Depth of line from surface
   Fline: Flowline
   Year_Const: Year of construction
   Comments: Notation of fixture's function
      Oleander = standard collection fixture
      Interceptor = large-size interceptor fixture
   RimElev: Rim elevation

Tools used

ArcGIS Pro:
- Select Layer By Attribute
- Append tool

As before, it is a good idea to make a field map or a diagram that shows which fields in the new feature class will be matched with which fields from the source data.

Create a field map

The field map will be a list of all the fields from your new data structure with their counterparts listed from the source data.

1. Print a copy of the completed tables worksheet from the geodatabase design forms you made for sewer lines in tutorial 1-2. On a separate sheet, copy the list of the fields you have designed. Compare these fields with the field names and descriptions for the SewerLineSource.shp file. On the right of each field from the SewerLineSource layer, write the name of the source field for that data and try to match the field description. Write your own first, and then check it against the figure.

Output Field Name	Source Field Name from Shapefile
Pipe_Size	Psize
Material	Material
Year_Built	Year_Const
Description	Comments

Load the data

Remember that the data will also be split into the feature classes SewerLines and Interceptors, on the basis of the contents of the Comments field. You will do this categorization with a query in the Append tool.

1. Start ArcGIS Pro, and open the project Tutorial 3-2.

2. As you did in tutorial 3-1, create a new group layer in the Oleander Utilities map, and add the SewerLineSource.shp and SewerNodeSource.shp files from Folders > Tutorial 3-2.

3. Find and open the Append tool.

4. **Set Input Datasets to Source Data Group Layer\SewerLineSource, and set Target Dataset to Utility Data Group\SewerLines. Build the expression** Where MATERIAL is equal to PVC.

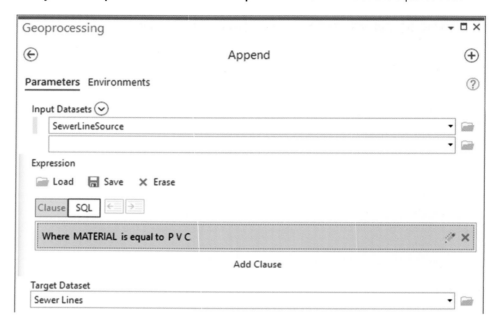

This clause lets you set the restriction to a specific material type, but you must also make sure that the features you select are owned by the City of Oleander. You can do this by adding a second clause to the expression using the Comments field.

5. **Click the Add Clause button. Build the expression** And Comments is equal to Oleander.

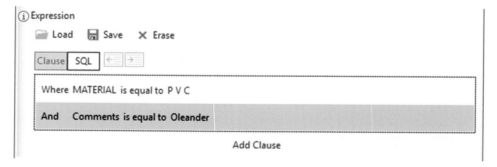

6. **Set Schema Type to "Use the Field Map to reconcile schema differences."**

7. **Set the field mapping according to the field map shown earlier, under "Create a field map."**

Note that Material was already mapped because the names for both the Output and Source fields were identical.

8. **Set Subtype to PVC. Click the Run button when everything is set.**

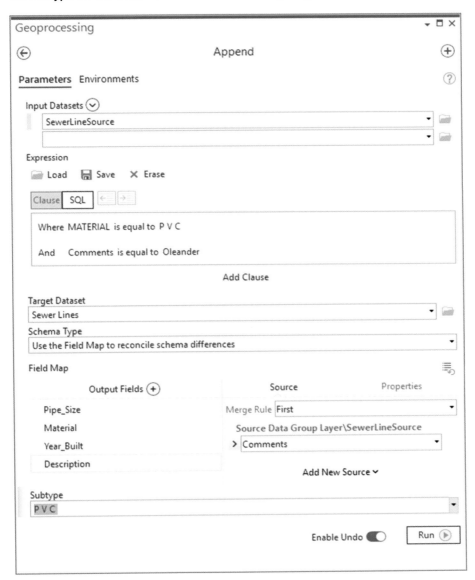

This process has loaded all the pipes that belong to Oleander and that are made of PVC. There are several more material types to load, so the process must be repeated for each additional type.

YOUR TURN

The load process for the next three set of pipes will be the same, except for two things—the Material query and the subtype. Change the Material query to read: Where Material is equal to HDPE and the subtype to HDPE. Leave the Comments query the same. You might want to enable the Undo switch in case you make a mistake along the way.

Repeat the data loading process for the rest of the material types:
CONC (RCCP), Clay (VC).

There is an additional subtype of DI for Ductile Iron, but there are no existing pipes of this material, so you do not need to run the load process for this subtype.

Review the geography and attribute table of the SewerLines feature class. If it is completed correctly, you will see all the data in both the attribute table and the map.

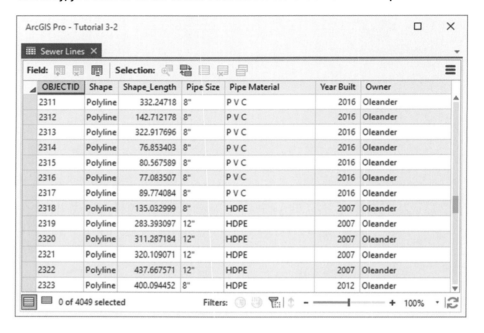

Now, you have loaded all the features for the SewerLines feature class, but the sewer line source data still has features that must be accommodated. That is the data for the interceptors.

All the interceptor data will go into one feature class without being divided into subtypes. Because of this provision, the Append tool will be run only once to put that data into the Interceptors feature class.

9. Open the Append tool. Set Input Datasets to Source Data Group Layer\SewerLineSource, and set Target Dataset to Utility Data Group Layer\Interceptors.

10. Set an expression to get only the lines where the Comments field equals T.R.A. Interceptor.

11. Change Schema Type to "Use the Field Map to reconcile schema differences," and set up the field mapping as before.

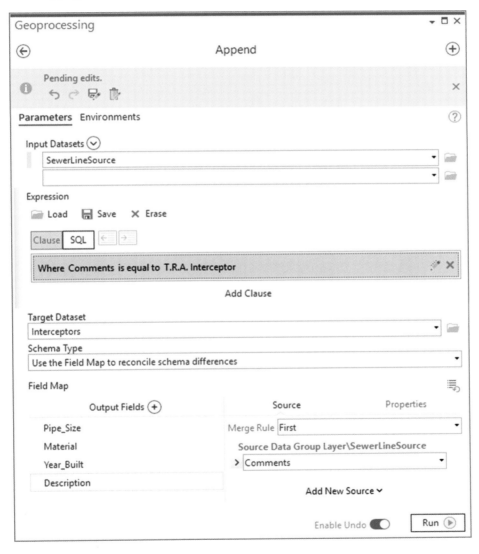

12. When all the parameters are set, run the tool.

Examine the results

1. Review the geography and attribute table of the Interceptors feature class. It should show just the larger lines that were selected with the query.

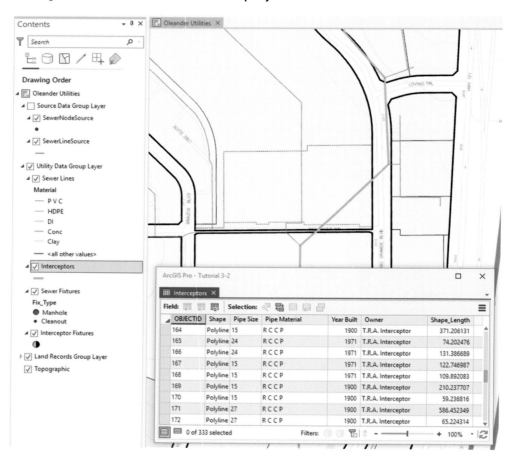

Load the point data

The loading process should be familiar by now, since it has a built-in repetitive nature, so the instructions to complete the process are mere guidelines. You will need to decide the tools and processes to use. Start by making a field map to determine what fields from the source data will match the new SewerFixtures feature class in the Wastewater feature dataset. The data description of manholes and cleanouts at the start of this tutorial will help.

1. Make a copy of the tables form from the geodatabase design forms for the SewerFixtures feature class in tutorial 1-2. On the right of each target field, write the name of the source field for that data. Write your own first, and then check it against the figure.

Output Field Name	Source Field Name from Shapefile
Fix_Type	SYMB
Flowline	Fline
Rim_Elev	RimElev
Depth	Depth
Year_Built	Year_Const
Description	Comments

The next task is to load the point data representing the sewer line fixtures into the SewerFixtures feature class. This process will again involve loading the data into individual subtypes based on the SYMB field. The following list from the previous data description will remind you of which values will be used:

    11 = manhole
    15 = cleanout

2. Remember to load only the features that belong to Oleander. Hint: Use a query when you load. Use the Append tool to put the features into the destination feature class.

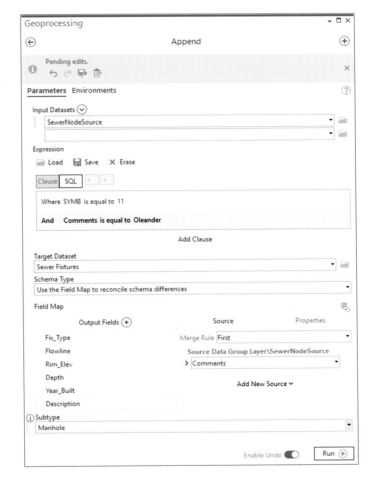

When the process is finished, you should have only the manhole features for Oleander loaded into the SewerFixtures feature class.

YOUR TURN
There's one more subtype to load: cleanouts. Repeat the load process, and load only the cleanouts for Oleander into the SewerFixtures feature class. Remember to use a query when loading so that unwanted data doesn't go into this subtype.

This addition should now bring the total features to 3,819 and should be all the Oleander data. You're closing in on the finish line with only one more task to complete. The Interceptor fixtures will all go into the InterceptorFix feature class without any subtype divisions.

YOUR TURN
Find the InterceptorFix feature class in the area where you've been working, and load the Interceptor fixtures from the SewerNodeSource shapefile. Use the query Comments is Equal to TRA Interceptor to find the desired features. Hint: Everything in this layer is a manhole, so there is no subtype.

Check the results

1. Review the geography and table for the SewerFixtures and InterceptorFix feature classes, respectively.

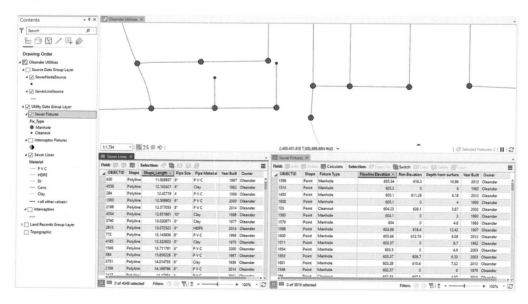

Chapter 3 | Populating and sharing a geodatabase   103

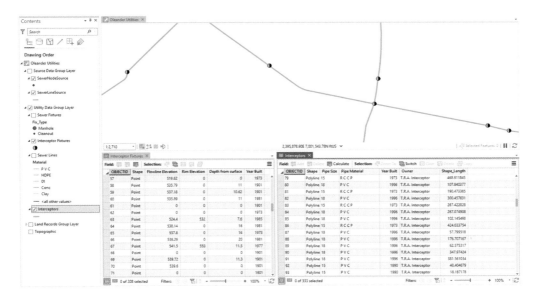

The import process is finally complete. As advertised, it's repetitive and tedious. It is important to pay attention to the minute details of the process to get through it cleanly and correctly. Otherwise, you'll have to delete what you've loaded into the geodatabase and start all over.

Exercise 3-2

The tutorial showed how to use the geoprocessing tools to selectively add data to an existing geodatabase.
    In this exercise, you will repeat the process and load data into the storm drain geodatabase you created in exercise 2-2. This geodatabase will include both the lines and points that represent the storm drain pipes and their fixtures.
- Start ArcGIS Pro. Add a feature dataset to the Utility_Data geodatabase, named according to your design forms.
- Investigate the fields in the Storm_Fix_Source.shp file in the Data folder, and match them to the fields in your new polygon feature class. Repeat with the Storm_Line_Source.shp file and the new linear feature class.
- Make a list of the fields in your destination feature classes and match them to the fields in the source subtypes before you start the data loading.
- Make a new map in ArcGIS Pro for the storm drain–related data.
- Use the Select Layer By Attribute and Append tools to load the storm drain data into the subtypes.

- The fixture codes are as follows:
    - 106 = "Y" inlet
    - 110 = beehive inlet
    - 101 = curb inlet
    - 102 = grate inlet
    - 109 = headwall
    - 104 = junction box
    - 107 = junction box/manhole
    - 111 = manhole (personnel access port)
    - 108 = outfall

Hint: Since these codes are already stored in an integer field in the source data, you will not have to load them individually into the subtypes. Once they are loaded, ArcGIS Pro will see these codes as correct subtype values.
- When completed, test the data integrity rules.

WHAT TO TURN IN

If you are working in a classroom setting with an instructor, you may be required to submit a screenshot of the geodatabase you loaded in tutorial 3-2:
- Screenshot image of the geodatabase previews from:
    - Tutorial 3-2 (geography and table)
    - Exercise 3-2 (geography and table)

Review

The geoprocessing tools Select Layer By Attribute and Append provide an easy, guided approach to importing geographic features into the geodatabase. As you can see from exercise 3-2, importing point and line features into the geodatabase uses the same process that you completed for polygons in exercise 3-1. Again, a well-thought-out approach to importing features from a shapefile into your geodatabase includes developing worksheets that document all your decisions during the import of your data.

Although the shapefile containing the sewer line data included all the different types of sewer lines in one shapefile, the import process allowed you to separate the sewer lines into two different feature classes according to the owner of the sewer line. To complete this process most effectively, you used the Select Layer By Attribute tool to query the Comments data field and appropriately isolate each of the sewer lines to load your data into new feature classes for sewer lines and interceptor lines. The import process also allowed you to further segregate the data you imported into existing subtypes in the geodatabase by populating the sewer line subtypes with the appropriate material for each pipe. For the point data, it let you delineate whether the subtype of each sewer node was a manhole or a cleanout.

One thing that should be evident is the importance of understanding your data. Familiarity with the data that is being imported as well as knowledge of how that data will be represented and manipulated inside the geodatabase is critical to your success. Exercises 3-1 and 3-2 provided examples of a real-life scenario in which data was imported from shapefiles that each stored data differently. Successful importing of the data required knowledge not only of how it was stored and managed in the shapefile, but also of how you wanted the data to be stored and managed in your geodatabase. In the real world, you will have to conduct the same logical steps to properly load source data into target data.

STUDY QUESTIONS

1. What is a load procedure, and why is it important for you to develop and follow one? What knowledge is required to successfully develop and execute the load procedure?
2. When using the geoprocessing tools, what options in the Select Layer By Attribute and Append tools must you ensure are set correctly before objects are loaded into a subtype?
3. What is the best way to quickly determine whether the import procedure was correctly performed?

Other study topics

Search for these key phrases in ArcGIS Pro Help for further reading:
1. Write a query in the query builder
2. Import Data

Chapter 4

Extending data formats

Once you have created and populated a geodatabase, there are many ways to extend the format. These methods include using the data in an online application, or even converting it to a 3D format. The basics of the data remain unchanged, but the techniques to edit and manage the data will change to accommodate the new format you use. This change may be minor, such as editing in 3D, which uses similar tools as editing in 2D but requires more considerations of the added dimension. Sharing your data online will change not only how it can be edited but also its characteristics, such as the types of symbology and annotation that are available.

Tutorial 4-1: Putting your data online

One of the best ways to share your data with others is by adding it to your ArcGIS Online account or Portal for ArcGIS in ArcGIS Enterprise. In that environment, you can control access to the datasets, allow editing, and easily use the data in web-based apps. There are also several configurations of data storage to consider, which will depend on your intended use.

LEARNING OBJECTIVES
- Share data
- Work with online data
- Build online web maps
- Manage access to online data

Introduction

Before sharing data online, it's important to know what storage components are available in ArcGIS Online and Portal for ArcGIS, and how your ArcGIS Pro project components relate to them. In your project, you will have layers compiled into a map. Each layer has properties set, and the underlying databases may have subtypes and domains. The map will have a spatial reference and extents set and will contain all the symbolized layers. It will have a name and may contain metadata, which could include tags, a summary, a description, a credit line, and any usage limitations. This hierarchy of layers and maps will be the basis for how you share data online.

When data is shared, it can become one of several components in ArcGIS Online. The simplest is the web layer, also identified as a hosted feature layer. When the web layer is created, it gets a companion service definition, which contains configuration data about the feature layer and is managed by ArcGIS Online. A feature layer can contain a single ArcGIS Pro shared map layer, or it may contain several map layers. Storing several map layers in a feature layer is a good way to keep like datasets together, even if they are of different geometries. For example, you may have police layers that include point data from phone records and linear data from a GPS tracker. When you share these layers in ArcGIS Online, you could put the two map layers into a single feature layer, making it easy and convenient to use in your ArcGIS Online apps.

The other ArcGIS Online component that you can use to share data is a web map. A web map is a map view used to display feature layers. In the map view, you can control the symbology, set up a pop-up box, control the transparency, and more. Feature layers must be placed in a web map for them to be displayed online. For more information on the components of ArcGIS Online, refer to the ArcGIS Online help.

Sharing a single map layer is simple. In the Contents pane, right-click the layer, and click Sharing > Share As Web Layer. All the parameters of the map layer are transferred to the feature layer, even definition queries. There are some limitations on symbology, but most symbols will bear a close resemblance to the ArcGIS Pro symbology.

To make a multiple-file feature layer with just a few of the layers from the Contents list, simply select more than one map layer before starting the sharing process. All the selected map layers will be placed into a single feature layer in ArcGIS Online. However, if you want all the layers to be placed into the web layer, you don't need to select them. Instead, on the ribbon, go to the Share tab, and click Web Layer > Publish Web Layer. Using this route will add all the layers in the map into a single web layer.

One of the most complex forms of sharing data is to share an entire map from a project as a web map. When an entire map is shared, a multifile feature layer will be created that contains all the layers in the map as well as a web map to contain them. The web map will have all the layers in it, each one symbolized to match the project map. The web map is then ready to include in map applications.

There are a few caveats when sharing layers. First, a full copy of the data behind the layers will be copied and stored online. There will not be a link between the layers, so editing either the desktop data or the online data will not alter the other dataset. You can, however, add the web data into a desktop map and edit the online file. Just be aware that sharing means making a copy of the data.

Next is the use of definition queries. If your project layer has a definition query when it is shared, the entire dataset is still copied to online storage and the definition query will still be enforced. If you remove the definition query, all the data will be present.

And finally, the spatial reference of the project map that you share must be WGS 1984 Web Mercator Auxiliary Sphere. If it is not, the projection will be changed automatically when you share the data. Use cf a standard projection places a common denominator on all your shared data, making it easy to include in multiple online applications.

Sharing your datasets

Scenario

The new City of Oleander Public Works director would like to see more of the utility data presented in online apps—something you've been pushing for. The department is budgeting to purchase tablet computers for all the crews and wants to be able to use them for inspections, line locates, information searches, and more. For a pilot study, you want to serve some of the wastewater data for them to try out and make it editable so that you can make changes as field reports come in. The data has the subtypes and domains you created, which will help maintain data integrity, even in the online environment. Because this project is just a pilot study, you will make two folders for the online data. One will have the project layers as single and multifile feature layers, and one will have the project map shared as a web map.

Data

The project contains clipped versions of several of the Oleander data layers you built earlier in this book. The dataset was clipped to reduce the time it will take to complete this tutorial and to minimize the impact on your ArcGIS Online data storage points.

The wastewater data includes a point and a line layer, each with the data integrity rules designed earlier. They are symbolized with generic point and line symbols. The property data includes the land-use layer and the parcel boundary layer. Note that a transparency has been applied to the land-use layer.

Tools used

ArcGIS Pro:
- Share As: Publish Web Layer
- Share As: Web Map

ArcGIS Online:
- Feature Layer: Item Details
- Web Map: Item Details
- Create Web Application

Preparing the data for sharing

1. Start ArcGIS Pro, and open the project file Tutorial 4-1.aprx.

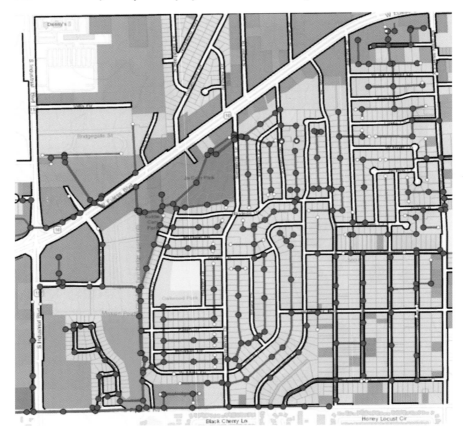

Before continuing, make sure that your ArcGIS Online organizational account has permission to publish data. This operation may consume some credits for storage, depending on how much data you are currently storing.

You will set the metadata for the Lot Boundaries project layer, and then make it into a single feature layer. The metadata that you create here will be transferred with the file to ArcGIS Online and become the data's name and description.

2. Right-click the Lot Boundaries layer, and open the Layer Properties pane. Click the Metadata tab.

3. **Change the metadata source at the top of the pane to "Layer has its own metadata." Name the layer** Lot Boundaries, **and add the tag** Oleander Public Works. **Then provide a summary and description. When you're finished, click OK.**

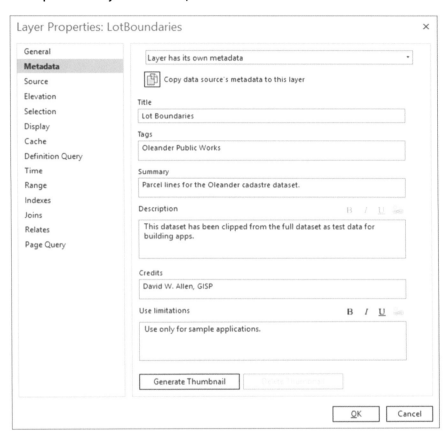

4. **Right-click the layer in the Contents pane, and click Sharing > Share As Web Layer to start the publishing process.**

5. **The Share As Web Layer pane opens. Note that the metadata you created earlier is passed into the pane.**

The location shown is the root folder of your ArcGIS Online account, which will be different from the one shown in the figure. You can make a new folder within this root folder to store the data, which will help keep it separated. For this test data, you will make a new folder.

6. **Click the Browse button to display the root folder. At the top of the pane, click New Item > New Folder. Name the new folder** Public Works Test 1, **and click Save.**

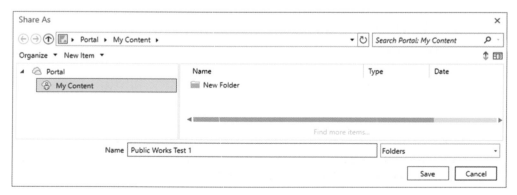

Before you attempt the sharing process, it is best practice to run the Analyze tool to make sure all aspects of the data are ready to publish. If there are any critical errors, correct them before proceeding.

7. **At the bottom of the dialog box, click the Analyze button.**

An error is displayed, and after reviewing the codes and reading the help, you can determine that the error isn't fatal. It involves a component of the layer that will be created automatically when the feature service is created. These are common errors and will appear on every layer you publish. You can ignore these types of errors when you publish data because they will resolve themselves.

8. **Click Publish. When the process completes, close the Share As Web Layer pane.**

9. **Log in to your ArcGIS Online account, and click the Content tab.**

10. **Look for the newly created Public Works Test 1 folder and open it.**

Note that two items were created. One is the hosted feature layer you were expecting, and the other is the service definition file that manages the ArcGIS Online components of the layer automatically. Both items must exist for the layer to be usable, so don't delete anything.

YOUR TURN

Go back to your ArcGIS Pro project and find the LandUse layer. Write metadata for the layer, and then share it as a feature layer to the Public Works Test 1 folder. Even though you know to expect one of the errors, it is still a good idea to analyze the data before sharing it in case there are other factors that you do not expect. When the layer has been shared, check your ArcGIS Online account, and make sure it is placed into the correct folder and has the correct components. Note that the large number of colors used here could not be set in ArcGIS Online but can be set in ArcGIS Pro before sharing the layer.

Share multiple layers

Now that those layers are in your ArcGIS Online content, you will need to add the wastewater data. This data consists of two project layers, but since they are always used together, you should make them a single feature layer in ArcGIS Online. The setup is similar, and just a small difference in the process will produce the desired results.

METADATA

An important part of sharing your data is to include good metadata. It not only can explain the origin of the data, but it can also give an idea of when it should or shouldn't be used for analysis. But why should you set up your metadata before sharing the layers? Once the layers are shared, it is much more difficult to create metadata. In the case of a feature layer that has two project layers, the new feature layer will have metadata that is unique from the layers you add to it. In this example, both the sewer lines and sewer fixtures will have metadata, and the new feature layer that contains them can also have metadata. It can be complex to set up all this metadata in ArcGIS Online, which makes it important to write the metadata first.

1. Edit the metadata for both the Sewer Lines and Sewer Fixtures project layers. Give them the tag of Oleander, and in the description, explain that the data has been clipped to a subset for demonstration purposes.

2. **Select both layers in the Contents pane, and then right-click one of the layers and click Sharing > Share As Web Layer.**

   Note the title of the sharing pane: "Sharing selected layers as a web layer." This pane also lets you assign metadata to the combined feature layer in addition to the metadata that comes with each project layer.

3. **Name the layer** Wastewater Layers**, and note in the summary which project layers are included. Set the folder to Public Works Test 1.**

4. **Click Analyze. You will get the same error message as before, but it can be ignored.**

5. **Click Publish.**

6. Open your ArcGIS Online account, and look in the Public Works Test 1 folder for the new layers. Note that only a single new item has been added. Click the Options button on the new feature layer, and then click View item details.

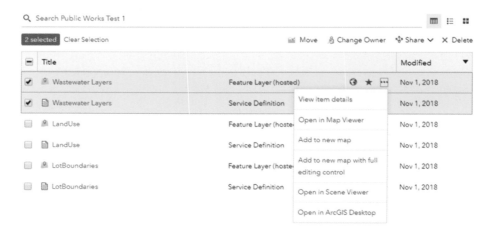

The layer details will show that both project layers were added as individual layers within this web layer.

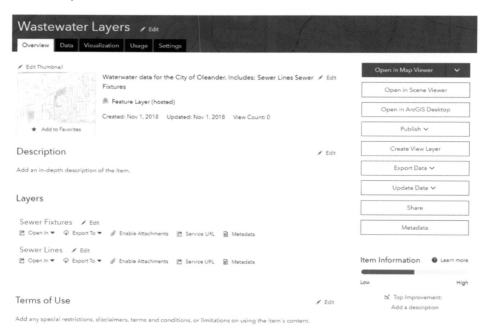

At this point, the pair of layers can be added to web maps or web apps as a single piece, but it will represent both the line and point features of the wastewater system.

Create a web map

Another aspect of data sharing that you can explore with these layers is making a web map out of the project map you have. The process here is also straightforward. On the Share tab is a button to create a web map. As stated before, this process will create a multilayer feature layer in addition to a new web map with the layers included. One change that you will do on this example is to make the data editable. Making it editable can be done after the data is shared, but it is much easier to do as part of the sharing process.

1. **In the ArcGIS Pro project, go to the Share tab, and in the Share As group, click Web Map. Don't worry about any layers that might or might not be selected because the entire map will be shared.**

2. **The Share As Web Map pane will look familiar. Under Location, click Browse and create a new folder named** Public Works Test 2.

3. **In the center of the pane, find the Select a Configuration line. Click the arrow and select Editable (an exploratory map with editing enabled).**

4. Click Analyze. This time a fatal error is reported. The map's projection is not the required Web Mercator as stated before. Hover the mouse at the end of the error message to expose the Options button, and click it.

5. In the Options pop-up menu, click Update Map To Use Basemap's Coordinate System. The error will turn green to show that the error has been resolved. Click Share. When it is finished, close the Share As Web Map pane, click Save, and close the ArcGIS Pro project.

6. Open your ArcGIS Online account, and look in the Public Works Test 2 folder. You will see a multi-layer feature layer and a web map.

Did you notice a missing layer? The Topographic layer that was in the project is not brought across as a web layer. The Share tool recognized it as a raster basemap and did not carry it across. So in your own maps, you will not have to worry about any of the basemap layers being stored in your ArcGIS Online account. You will find, however, that the web map that was created will have these web layers, and the reference to the Topographic basemap that was being used will be automatically added back to the map.

7. **Examine the item details for the layer, and then open the web map in Map Viewer. Notice that all the layers are included, and the default symbology is used.**

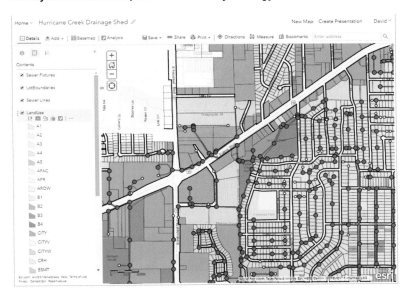

8. **On the toolbar, click Edit to open the Add Features toolbar.**

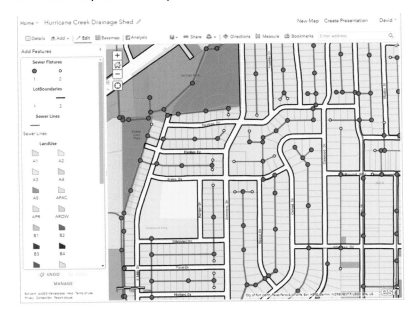

9. Click the manhole symbol (large red dot), and click in the map somewhere to add a new feature. The attributes box will open. Note that the domain is applied to the Sewer Fixtures type, and the defaults from the database design are applied.

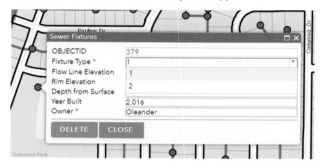

10. If you want to try other features to test the data integrity rules, recall that the Lot Boundaries layer has a domain on the Linecode field, and the Land Use layer has a domain on the Use Code field. After examining the data, save the web map and close ArcGIS Online.

The web map can be dropped straight into a web app, or even opened in ArcExplorer. Although this book doesn't cover using ArcGIS Online, you might want to experiment with this clipped dataset and try different templates and applications.

Exercise 4-1

There is also some storm drain data that should be added to the Public Works Test 2 folder in your ArcGIS Online account. The project has a second map named Blessing Branch Drainage Shed that contains this data. The goal will be to add this data into the web map that you created in the tutorial. Examine the data and determine the best sharing process to get the data into the existing web map.
- Start ArcGIS Pro, and open Tutorial 4-1.aprx, if necessary.
- Open the Blessing Branch Drainage Shed map.
- Review the data in the Contents pane, and add metadata as needed.
- Determine the best method for sharing this data to the Public Works Test 2 folder in your ArcGIS Online account.
- Share the data, and add to the existing web map.

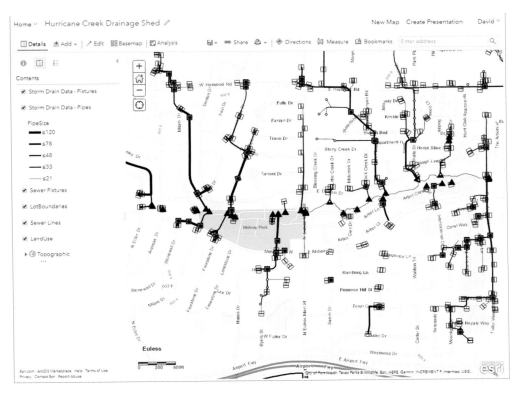

- Save the results with your ArcGIS Pro project.

WHAT TO TURN IN

If you are working in a classroom setting with an instructor, you may be required to submit the maps you created in tutorial 4-1.
- Screenshot images of the web layers you created in
  - Tutorial 4-1
  - Exercise 4-1

Review

Sharing your data through your ArcGIS Online account or Portal for ArcGIS makes it available for many online uses. The process is simple using just a few button clicks. The important things to remember are that the metadata must be complete, the spatial reference for any web maps must be WGS 1984 Web Mercator, and you may be limited in some of the symbology you can use. The good news is that the data integrity rules you designed for the data are enforced on the web data, so any editing that is done online will follow your rules.

It's possible to share individual layers in a project or in an entire symbolized map. These layers are re-created in your ArcGIS Online account as web layers and web maps. Once there, they are available for use in web apps. It is also desirable to store these layers in a separate folder so that they are easily identified and found when needed.

STUDY QUESTIONS

1. Why is adding metadata important?
2. Compare the capabilities of a layer in a project versus a web layer in ArcGIS Online.
3. Describe how the process and results differ from sharing a single layer to sharing a map.

Other study topics

Search for these key phrases in ArcGIS Pro Help for further reading:
1. Share with ArcGIS Pro
2. Introduction to sharing web layers

Tutorial 4-2: Creating 3D scenes

The data created so far is two-dimensional, since it takes the 3D Analyst extension to create true 3D data. However, if you plan ahead, you can present much of your 2D data in a 3D view. These views are called *scenes*, and this tutorial will take a brief tour of creating scenes.

LEARNING OBJECTIVES

- Create a new scene
- Prepare 2D data for use in a 3D scene
- Work with elevation data
- Set 3D display parameters
- Use 3D navigation tools

Introduction

When new feature classes are created, there is a prompt to allow the data to store z-values. If this prompt is selected, and it is the default, the Shape field of the data can store a 3D value. Using the 3D Analyst extension, you can put the data into a true 3D view and perform analysis on it, intersect it with other 3D data, or even use it as the basis for overlaying 2D data.

Even without the 3D Analyst extension, there are ways to present your data in a 3D view, or scene. The scene can get an elevation basemap from ArcGIS Online to use as a reference, much in the same way that a project's map view gets a 2D topographic map from ArcGIS Online. Once the basemap reference elevation is set, the features can be shown in relation to the basemap. Each element in the feature class must have an elevation value (or be marked as No Data). This elevation can be in reference to an absolute elevation measured from sea level or referenced to the ground. For example, if you collected elevations from a weather balloon using an altimeter, the reference would be in mean feet above sea level. And if you were building radio transmission towers in heights of 200, 250, or 300 feet, they would be referenced from ground elevation. You would add the height of the tower to the ground elevation to find the top of the tower. You can even reference the ground in a negative direction. For instance, you may be mapping the depth of gas wells, which could be 2,000 to 4,000 feet deep. The numbers would be recorded as a negative number, meaning that you would start at ground elevation and subtract the depth of the well.

It's also interesting to note that the values used for the 3D representation don't have to represent elevation. It would be possible to use a value and add it to the ground elevation to show greater and lesser magnitudes. For instance, the same wells showing a negative depth value could also show a positive value of cubic meters of gas produced. Be careful, though, because the units will always be measured in map units. So, 200,000 cubic meters of gas well production would be interpreted as 200,000 feet in elevation. When using these types of values, it may be necessary to add an expression to the elevation value so that values in the hundreds of thousands don't project vertically out to the moon. Limiting these values can be done easily when setting the elevation field using the optional expression builder.

Putting Oleander in 3D

Scenario

The City of Oleander just got new building footprint data and would like to create a 3D scene in ArcGIS Pro. Then you also want to show a representation of manhole depth of the wastewater data so that the public works crews will have an idea of how deep they would dig for any repair job. Repairs at shallow manholes can be done without shoring, but for deeper excavations, they may need larger equipment, OSHA-approved shoring, and a 3D representation to help them plan what equipment will be necessary.

Data

The datasets include the building footprint data and the wastewater manhole data. The building data has a field named Height that represents the height of the building. The wastewater data has a field named Depth from Surface that is the distance from ground level to the bottom of the manholes.

Tools used

ArcGIS Pro:
- Create a Scene
- Set 3D elevations

Examining the data

1. **Start ArcGIS Pro, and open the project file Tutorial 4-2.aprx.**

You can see that the layers in this 2D display include the building footprints and the sewer fixtures (manholes). You will need to check that both files have a field that can be used to represent an elevation.

2. **Open the attribute table for Building Footprints. Look for the field Height, and scan through the values.**

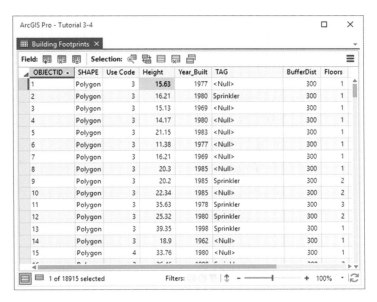

These values look as if they will work. Some values are zero, but they represent buildings that were razed but the slabs were left. Note that the values are the measure from ground level to the top of the roof. Note the field Floors. If a Height of Building field was not available, it still would be possible to do a rough estimate by multiplying the number of floors by 10.

3. **Open the attribute table of the Sewer Fixtures layer.**

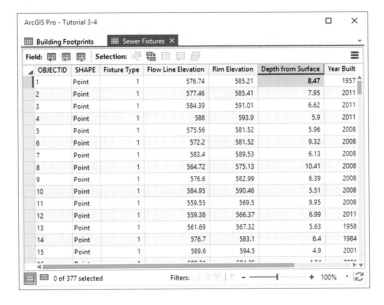

One field represents the depth of the manhole, but it is a positive value. If it was used as is, the manholes would protrude from the ground rather than extend below the surface. The simple fix is to make the depth value negative so that the measurements will be subtracted from ground level.

4. **Right-click the Depth from Surface field, and click Calculate Field. Set the expression to** Depth * -1. **Click Run.**

5. **Verify that the values are now negative, and close both attribute tables.**

Convert the map to a scene

The map can be converted to either a local scene or a global scene. The difference is the location of the viewer's perspective. With a global scene, the viewer is in space looking back at a globe. This type of scene is best for data that covers a large part, or all, of the globe. The viewpoint of a local scene is typically a few miles above the surface, displaying a localized area. This type of scene is best for small study areas, which is what you will use for the small area of Oleander.

1. **On the ribbon, go to the View tab.**

2. **In the View group, click Convert > To Local Scene.**

A new scene is created named Downtown Oleander_3D. Note in the Contents pane that there are two data groups: one for 3D Layers and one for 2D Layers. Notice also that an Elevation Surfaces group was added, and the default World Elevation 3D layer was added, which becomes the elevation reference for the rest of the data. These are the building blocks of the 3D display.

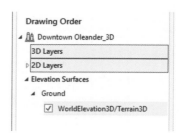

Before layers can be represented in 3D, they must be dragged into the 3D Layers group. Then the elevation field can be identified.

3. **Drag the Building Footprints layer from the 2D Layers group to the 3D Layers group.**

4. **With the Building Footprints layer selected, go to the Appearance tab and find the Extrusion group.**

The tools contain several extrusion methods, but some of them require a true 3D dataset. Because this is not a true 3D dataset, it has no z-values, so the first two options won't work. The setting you will use is Base Height, since this setting will extrude the buildings up from the ground. The setting of Absolute Height would establish the buildings above the surface of the ground. Note that if you have the 3D Analyst extension, you could run the Interpolate Shape tool to make the dataset z-aware.

5. **Click Type, and select Base Height.**

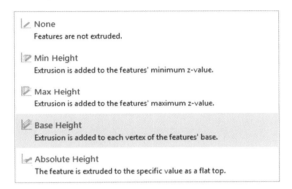

6. **Set Field to Height.**

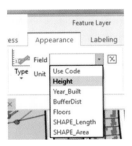

Note that on the right of the Field box is the Expression button. You can build an expression to get the elevation. As mentioned before, an expression can be built to use the value of the Floors field to calculate a roof height.

Setting the Extrusion field establishes that the layer is laid over the elevation surface. This option sets only the base height of the layer but not the extruded height of the roofs. That is done in the layer's properties.

7. **Right-click the Building Footprints layer, and open the Properties pane.**

8. Go to Elevation, and verify the setting "On the ground." Click OK to close the Properties pane.

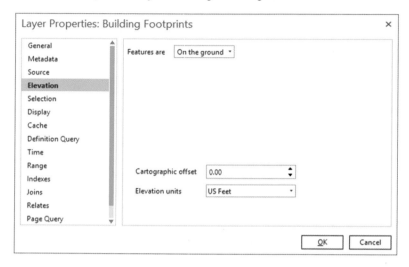

That's basically it. The buildings will now be shown in 3D. Use the On-Screen Navigator to control the view. This control is complex since it must navigate you through 3D space, so the best way to learn how to use it is to read the ArcGIS Pro Help page. (Hint: Search for "Using the On-Screen Navigator.")

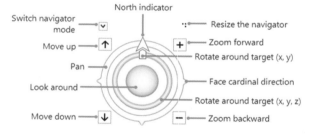

Chapter 4 | Extending data formats 129

9. Navigate around and view the building representations.

YOUR TURN

The extruded building data is shown as a single color, but you could also use any of the layer attributes to set a different symbology. You might try symbolizing the layer using the Tag field to show which buildings have a sprinkler system installed, or use the YearBuilt field to show the age of each structure. Or use the Use Code field to show the land-use code for each building.

Work with negative extrusion values

The buildings extruded to a positive height, and now you will work with the other data that extrudes in the negative direction. The data has already been recalculated to represent a depth below the surface, so you need to set only the extrusion and the layer properties.

1. Drag the Sewer Fixtures layer into the 3D Layers group, and place it just above the Building Footprints layer.

2. You may have navigated away from the extent of this data. To return to it, right-click the layer and click Zoom To Layer.

3. Go to the Appearance tab, and set the extrusion type to Base Height and the field to Depth from Surface.

4. Open the layer properties, and set the Elevation parameters to "On the ground," if necessary.

5. Finally, turn off the World Elevation 3D and Topographic layers. Because these are raster layers on the ground surface, they prevent you from seeing below the ground.

6. Navigate through the data to see the results. Each manhole will be extruded below the ground surface according to its recorded depth.

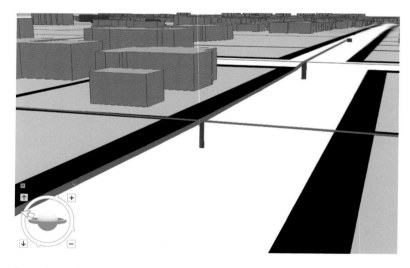

7. Save the project.

    If you are not continuing to the next exercise, exit ArcGIS Pro.

Exercise 4-2

Now that you've seen the subsurface wastewater fixtures, the public works director would like to see how they compare with the subsurface storm drain features. The layer has a field named Invert (Fixture Depth). Use it to build a 3D representation of the storm drain data. It may need to be turned into a negative value—see if you can do that with an expression.
- Start ArcGIS Pro, and open Tutorial 4-2.aprx, if necessary.
- Add the StormFixtures_Web data to the scene.
- Move the layer to the 3D Layers group.
- Set the layer properties for Elevation to display the features.
- Save the results with your ArcGIS Pro project.

WHAT TO TURN IN

If you are working in a classroom setting with an instructor, you may be required to submit the maps you created in tutorial 4-2.
- Screenshot images of:
  - Tutorial 4-2
  - Exercise 4-2

Review

This tutorial demonstrates the use of 2D data in a 3D environment. The project uses an online dataset as the basis for the ground, and all other data is established in relationship to this dataset. Note that both the 2D and 3D views of data can be viewed at the same time, and editing can take place in either view. For true 3D capabilities, however, the 3D Analyst extension is required. Viewing 2D in a 3D environment opens a new realm of using 3D data in analysis such as contour mapping, viewshed calculations, drainage flows, and more.

STUDY QUESTIONS

1. What is the basis for the scene elevations?
2. What is the difference between the base and absolute extrusion types?
3. What would you need to do to make a dataset z-aware?

Other study topics

Search for these key phrases in ArcGIS Pro Help for further reading:
1. 3D GIS terminology
2. Navigation in 3D
3. 3D effects—ArcGIS Pro
4. Extrude features to 3D symbology

Chapter 5

Working with features

The first few chapters were all about designing and creating the structure to hold your data, the geodatabase, and about how to bring datasets into your new schema. This chapter will explore ways to edit the data within the schema to take advantage of the best editing tools as well as the data integrity rules. When editing data, it's easy to fall into a rut using the same tools and techniques that have become familiar, but it's a good idea to explore all the tools to make sure that you are working with the best tools possible. Knowing what these tools are and how to use them, in conjunction with the standard editing tools, will help you improve your editing skills and save time when creating data.

Tutorial 5-1: Creating new features

ArcGIS Pro includes many techniques for creating new line features, including a variety of editing tools and COGO data entry tools. Using these tools, you can draw any level of accurate data in your project, whether it comes from rough diagrams or survey-quality plans. The key is matching the tools to the task.

LEARNING OBJECTIVES
- Set up group layers
- Organize the Contents pane
- Work with bookmarks
- Set map extents
- Create new features
- Use the context menu tools

Introduction

Tutorials 3-1 and 3-2 showed how to take existing data and import it into a carefully crafted geodatabase structure. However, not all the data you might need will already be created. Sometimes you will have to create the data yourself. ArcGIS Pro contains a large array of tools for creating points, lines, and polygons, and the trick to creating new features is to determine the correct tool to use. Sometimes creating new features is done with absolute coordinates, sometimes with survey data, and sometimes by describing a feature's location in relation to other features.

This tutorial will use a combination of tools to turn a rough schematic into GIS data.

Creating detailed features

Scenario

The Oleander Regional Transportation Authority (ORTA) has received the first draft of plans for a new rail line and commuter station in Oleander. It has provided a measured drawing, but it is not based on clean survey data. Instead, it is a collection of data from which you will need to reproduce the project as an overlay for the existing Oleander data.

The city planner would like to see the train track alignment overlaid on the parcels to study the potential impact of noise and pollution. He'd also like to see the location of the station and parking lots to determine any potential traffic problems. The results of his analysis will be presented to the Oleander City Council to determine the level at which the city may participate in the ORTA funding.

The drawing provided should be enough to draw the proposed location with adequate detail to do the preliminary study.

Data

Empty line and polygon feature classes are provided in the ORTA geodatabase, which you may use to draw the plan. Also provided are the Oleander parcel and lot boundaries layers to provide background reference.

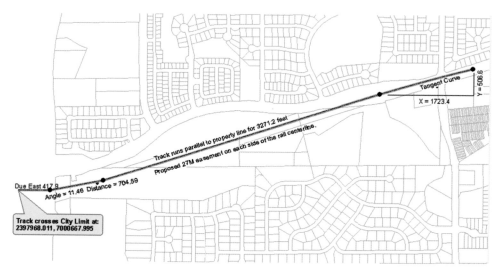

Tools used

ArcGIS Pro:
- Group layers
- Bookmarks
- Specify extent
- Line: Absolute X,Y,Z
- Line: Delta X,Y,Z
- Line: Direction/Length
- Line: Parallel
- Line: Length
- Tangent tool

Look at the diagram provided under "Data," which shows the proposed rail line and station. The first step is to draw the rail line. The data seems haphazard and disjointed, but with the right techniques, the rail line can be accurately drawn. Drawing it will involve using a combination of the standard editing and context menu tools.

Preparing the table of contents

1. **Start ArcGIS Pro, and open the project file Tutorial 5-1.aprx.**

   In this map, the Contents pane includes several layers. The map displays all the parcels and property boundaries located within the City of Oleander, the survey control monuments, and some empty feature classes that will contain the new data for the proposed rail facility. In addition, the data isn't being shown very clearly.

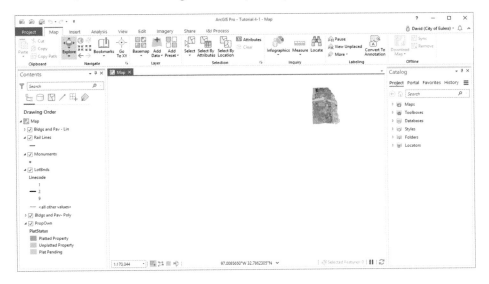

There's a lot going on in this map, and none of it looks good. Before you can begin to edit and add new features, you must do some housekeeping to make the map project easier to work with. The first task is to tidy up the list of layers in the Contents pane. The feature classes aren't in any order, and the names need to be more descriptive of the project. Remember that when a legend is added to the map, the names of the layers appear as they are in the layers list, and they will need to describe the features so that the reader can interpret them. For managing purposes, you'll change the name of the map frame to be more reflective of the project, add a new group layer to the layers list, and then arrange the layers in the group.

2. **Right-click the map frame name (Map) at the top of the layer list and click Properties. Click the General tab, and type the new name of** Proposed Rail Station. **Click OK.**

An easier and faster way to rename any map frame or layer is to click the name, pause slightly, move the mouse a tiny bit, and then click again; or highlight it and type F2. The name will be in Windows edit mode, and you can retype it. Be careful not to click too rapidly or it will be detected as a double-click and open the properties of the item. It might take some practice to get the hang of it.

Next, you will add a group layer to hold the basemap information. The group layer will contain the layers that are not being edited in this map, making them easier to manage.

3. **Right-click the map frame again, and click New Group Layer.**

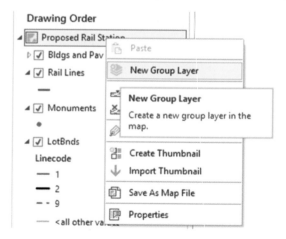

A group layer is a device to hold feature classes and allows common properties to be set for all the feature classes it contains. It also lets you control the visibility of all the layers it contains using a single mouse click. Because group layers look and act just like feature classes, it's best practice to always put the word *Group* in their name. These groups will also be shown in the legend, so they are useful in keeping similar layers together for the benefit of the viewer of the printed map output.

4. **Change the name of New Group Layer to** Property Base Group Layer. **You may decide whether to open the Properties dialog box and change the name, or try the Windows edit mode.**

   This new group should contain the following layers:
   PropOwn
   LotBnds
   Monument
   Move these layers around simply by dragging them into the group. The name of the group will highlight as you drop a layer onto it, and after the first layer is added, a horizontal bar appears to show where the layer will display.

5. **In the Contents pane, drag the PropOwn layer into the Property Base Group. Watch for the box to appear around the group name.**

   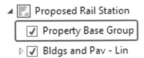

6. **Drag the Monuments and LotBnds layers into the Property Base Group. Note the line indicating where the layer will appear in the group.**

   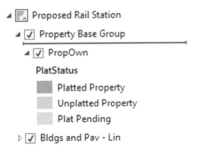

   The layer Monuments needs a better name. When the layer name appears in the legend, the viewer should be able to understand what data that layer contains, and right now it is too generic and nondescript.

7. **Change the name of the monument layer to** Survey Control Points. **Use either the Windows editor shortcut, or change it in Layer Properties. Also, change LotBnds to** Lot Boundaries **and PropOwn to** Property Ownership.

All this name changing makes a huge difference in how the layer list, and ultimately the legend, looks. But there are other benefits to the group layer.

8. **Right-click Property Base Group Layer, and click Properties. Make sure you are viewing the General tab.**

Under this tab, you can change the name and visibility of the group and set a visible scale range. Under Metadata, you can add a description or data credits. These parameters will be applied to all the layers in the group.

> **ADJUSTING SCALE FOR GROUP LAYERS**
> The visible scale ranges of the layers and the group layer can be manipulated to give precise control over everything in the group. For example, the group may contain three layers of data that are appropriate at only medium zoom scales, one layer of specific data that should be shown only at a close zoom, and annotation that should also be shown only when zoomed in tight. You could set an individual scale range for the two layers that require a tight zoom, and then set an overall scale range for the group that would turn off the layers it contains when zoomed out too far. This matrix of visible scale range settings will work together to turn on data when appropriate.

9. **Click the Metadata tab to examine parameters such as Tags, Summary, and Credits. Here, you can record valuable information about the data, such as what is shown in the figure. Click OK to close the dialog window.**

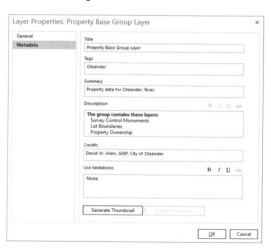

10. **With Property Base Group highlighted, click the contextual tab Layer > Appearance on the ribbon. Note the settings and tools.**

This context-driven menu allows you to make display settings for all the group's layers at once. The scale visibility controls are accessible here, too, but the others are controlled only here. The slider bar will set the transparency for all the layers in the group. Be careful if you have previously set a transparency level for any one of the layers, because this setting will reduce the transparency level even more by the percentage you set. ArcGIS Pro will simulate the transparency levels in the legend, making the output map easier to read and understand when transparencies are used.

YOUR TURN

Add another group layer for the North Oleander Station data. Move the two layers for buildings and pavement (lines and polys) and the layer for the rail lines to the new group. Change the name Bldgs and Pav – Lin to Buildings and Paving – Lines **and Bldgs and Pav – Poly to** Buildings and Paving – Polygons. **Your final layer list in the Contents pane will now be clear and more descriptive of your data.**

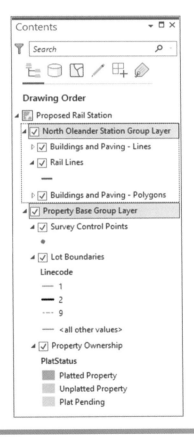

The group layers are a useful way to arrange the data to control some of the common properties across the layer list. Groups can also be nested, so that a group to contain basemap information might contain a group for Oleander data, a group for Tarrant County data, and a group for state of Texas data. This grouping will result in a clean, organized layer list and, as a result, an attractive legend. The symbol levels, which will be used later, can control the drawing order, even across groupings.

The last step in setting up your project is to establish which layers can or cannot be edited. Because the Survey Control Points, Lot Boundaries, and Property Ownership data should not change while editing the rail line, you will turn off editing for these layers.

11. **In the Contents pane, click the List By Editing tab. Then clear the check box next to the three layers in Property Base Group.**

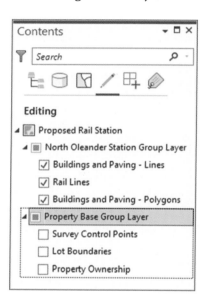

Work with bookmarks

The map is almost ready to begin editing, but first the area of interest must be located. The designers have previously zoomed in on the area of interest to study the site, and they have saved a bookmark for you.

1. **On the ribbon, go to the Map tab, and in the Navigate group, click Bookmarks. Click the Orta Site bookmark.**

The map is now zoomed in on the area of interest, and the Contents pane is tidy with the layers named appropriately. It's almost time to start editing.

Control default map extents

The last thing to change is the zoom extents of the Zoom to Full Extent button. When editing, you will be zooming in on small areas to see more detail, and then wanting to zoom out to the full extent of the study area to see the results. Although bookmarks are useful in this regard, the Full Extent button is much faster and more convenient. First, try out the button and see the results.

1. **On the Map tab, in the Navigate group, click Full Extent to zoom to the full extent.**

The map zooms out to a view seemingly from "outer space." No work could be done with any precision at this zoom level. You'll need to return to the previous zoom level, and set it as the new default zoom extent.

2. **On the Bookmarks menu, click Orta Site to return to the area of interest.**

3. **Open the properties of the map frame Proposed Rail Station, and click the Extent tab.**

4. **Change the setting to "Use a custom extent." Then under "Get extent from . . .", select "Current visible extent." Click OK to close the Map Properties box.**

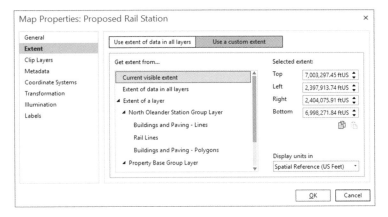

5. **Test the extent setting by panning to a new area, and then clicking the Full Extent tool.**

Setting the extents this way will prevent the outer-space view of your data from constantly appearing. By default, the extent of the full zoom is set to the largest extent of the layers in your map. If you happened to bring in some state or national data, your default zoom extent will be very large. After setting this parameter, you have better control over the extent and will save a lot of time traveling to and from outer space to see your data.

This preparation seemed like a lot of work just to set up the project, but it will pay dividends in the future in the following ways:

- You'll save time by not getting lost in far-out zooms.
- The data is easier to see and understand in group layers.
- Only the layers for the new rail system can be edited, protecting the other layers from accidental changes.
- You'll save time when you add a legend, because the data is already named and organized.
- You'll be able to turn whole sets of data on and off by the group layers.
- You can use a group properties dialog box to manage the layers included in a group.
- Anyone opening this map document will find it easier to understand the data.

And much more.

Creating line features

Now back to the editing tasks.

Begin by examining the alignment of the rail line, which will be drawn as a single line representing the centerline of the track. It begins at a known x,y coordinate and describes the location through various dimensions. These dimensions were taken from field observations, known rail standards, and other methodologies. Although it would have been easier if a single method was used, it doesn't always happen that way in the real world when initially planning a project.

To draw this line, you will use a variety of tools from the editing context menu. By combining several of them, you'll be able to draw the line.

1. **Zoom to the full extent, if necessary, and click the Edit tab. Then in the Features group, click Create.**

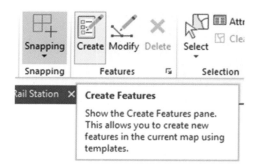

The Create Features pane opens, displaying the editable layers. To create new features, you will select the desired feature from this pane and use the construction tools to draw it. Note that the layers you cleared on the List By Editing tab do not appear in the Create Features pane, nor are they open for editing.

2. **In the Create Features template, click Rail Lines, and then click Line in the Construction Tools pane.**

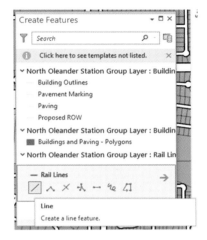

Notice that the layer names give a clear idea of what will be edited with each selection. Imagine if the Create Features template was to have a list such as Layer 1, New Features, Lines, and Old Polygons. Would you know which one to select? You can appreciate the value of giving your layers descriptive names.

The Construction Tools pane contains all the cursor-based tools available for drawing new features. On the right side of the pane is an arrow that will open other options. One option is to prepopulate attributes for features before they are drawn. For example, before drawing all the parcels in a new subdivision, you may want to prepopulate the values for the City and Subdivision Name fields rather than select and edit them later. For the rail lines, you can prepopulate the value for the line's description.

3. **Click the arrow on the right side of the Construction Tools pane.**

4. **Set the value for Description to** Proposed Oleander Rail Location. **Then click the arrow to return to the feature templates.**

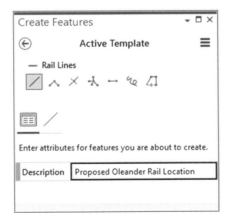

When you first opened the Create Features pane, you may have noticed a new tool bar that popped up at the bottom of your map. This ghost toolbar gives you quick access to the most commonly used construction tools, saving you time and mouse movements to continually go back to the main Create Features pane.

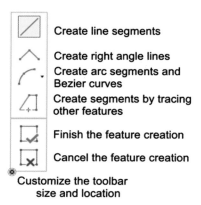

The toolbar contains an option to control its size and location. By hovering over the button in the lower-right corner, you will see the pop-up Configure Tool feedback options. Clicking the button allows you to choose one of three preset sizes and to display the toolbar on either the bottom, right, or left side of your map. This example shows the toolbar moved to the left of the screen, but for now keep it at the bottom.

As you move through these instructions, you will use many of the context menu construction tools. These tools are accessed by right-clicking in the map area. The context menu will change to suit the current process, and the figure shows a view of the context menu for line creation. Note that the ghost menu appears on the top of the dialog box, followed by a list of tools. Some tools are dimmed because the current context doesn't allow their use. They will become available as the situation changes.

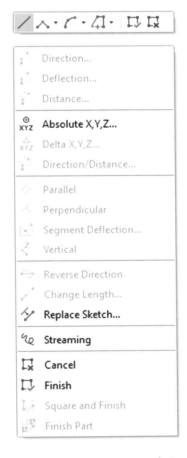

The first tool that you will use to create the proposed rail line is the Absolute X,Y,Z tool. On the diagram, the x- and y-coordinates of the beginning of the rail line are provided. These values are in projected map coordinates, which the Absolute X,Y,Z tool will convert to a location on your map.

5. Move the cursor anywhere in the display area of the map, right-click to expose the context menu, and click the Absolute X,Y,Z tool. Set the units to feet (ftUS), if necessary.

6. Enter the coordinates 2397968.011 for the x-value and 7000667.995 for the y-value, as was shown in the diagram under "Creating detailed features" at the beginning of the tutorial. Be careful not to press Enter until both values are correct. Hint: You can move between the two entry boxes by clicking the mouse or using the Tab key.

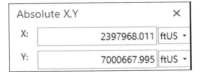

This setting will enter the first vertex of the new line, and you will see a red square showing the location in addition to the rubber band stretching to the cursor location. As you use more of the drawing functions, more vertices will be added to the line. The next tool to use is Delta X,Y,Z which uses the change in the x-direction and the change in the y-direction to locate the next vertex.

7. **Right-click the display area again, and from the context menu, click the Delta X,Y,Z tool.**

8. **Type 417.9 for the x-value and 0 for the y-value. Press Enter when both values are correct.**

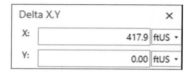

With these two points entered, the line will start to form a sketch. This temporary drawing, shown as a dashed line, will not become a feature in the target layer until you double-click at the last entry point, press F2, or click Finish to complete it. Note that the Finish command is available on the ghost menu and the context menu. The last location entered is always shown in red, with the rest of the locations shown in green.

If you make a mistake when entering a node in your line, move the cursor over the node, right-click to expose the context menu, and click Delete Vertex. Alternatively, you can click the Undo button or press Ctrl+Z.

Next, you'll use the Direction/Length tool. This tool will accept the angle of the line, along with the distance to locate the next vertex. Note its keyboard shortcut for future use.

9. **Expose the context menu again, and click Direction/Distance. Change the direction type to *P* for Polar Coordinate, and then enter the direction 11.46 and the distance of 704.59 (ft), and press Enter. This step adds another vertex and extends the line.**

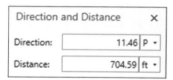

SETTING MAP UNITS

The direction value given in the sketch is in a polar angular measurement. Angles start at a single pole, or zero position, and are measured counterclockwise in degrees.

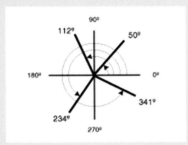

The various map units can be set on the fly as you edit, as you've been doing in this tutorial. But you can also set them before you start editing so that they will match the requirements of the project.

To set the units, open the Options dialog box on the Project tab and make the appropriate settings to any of the map units.

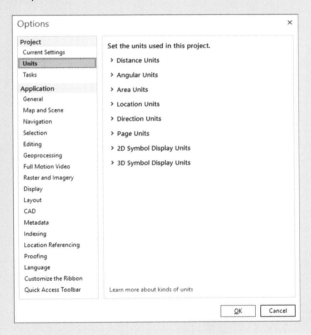

The next segment will be constructed using a combination of tools. The first will be a context menu selection to make the sketch parallel to an existing feature. The second will be a different context menu selection to set the length. Once the angle and distance values are given, ArcGIS Pro has enough information to construct the next segment and will extend the sketch.

For the Parallel function to work, you must identify for ArcGIS Pro which feature you want to get the angle from. You'll place the sketch cursor, shown as a crosshair, over the desired line and invoke the Parallel function from the context menu.

An existing property boundary that represents the edge of the railroad right-of-way will be used to set the angle of the new line.

10. **Move the cursor over the diagonal feature shown in the figure, identified as Lot Boundaries: Edge by the snap tip, and then right-click and click Parallel.**

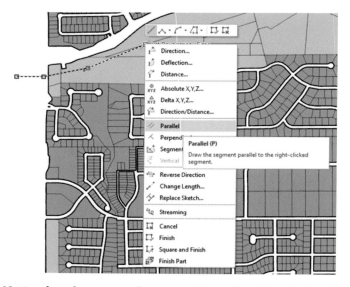

Notice that if you move the cursor, the angle of the sketch will remain parallel to the line you chose. If you had selected the wrong line, or did not want to use that angle, you could release this constraint by pressing Escape.

11. **Right-click and select Distance from the context menu. Type the distance as** 3271.2 **(ft). This portion of the sketch will be drawn, and you will be ready for the next location.**

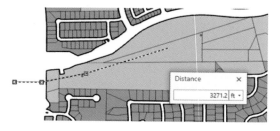

The final segment of the rail line curves off to the edge of the area of interest. This segment will require a different Sketch tool, Tangent Curve Segment, to draw the arc segment.

There are nine Sketch construction tools, as shown in step 10, and more information can be found in ArcGIS Pro Help by searching for "menu and toolbars for editing."

The Tangent Curve Segment tool will draw an arc at the tangent angle from the last sketch segment drawn. Using this tool with a length, you will be able to finish the construction.

12. **On the Feature Construction toolbar, click the arrow in the Arc Segment construction tool, and select the Tangent Curve Segment tool.**

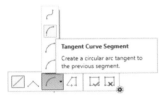

13. **Expose the context menu, and click Delta X,Y,Z. Type the values of** 1723.4 **(ftUS) for X and** 506.6 **(ftUS) for Y, and press Enter. Finally, click Finish using one of the methods discussed earlier to complete the creation of the new feature.**

This line creation seems like a long process, but as you become more familiar with the Feature Construction tools, things will go smoother. The new rail line is drawn, but the symbology isn't too appealing. So, you'll set it as a double-line railroad symbol.

14. **In the Contents pane, move to the List By Drawing Order tab and expand the view of symbology for the Rail Lines layer, if necessary. Click the current symbol to open the Symbology pane. In the search line, type** railroad, **and press Enter. Several railroad symbols are found. Select the symbol that displays the track in black, and click OK. Then close the Symbology pane.**

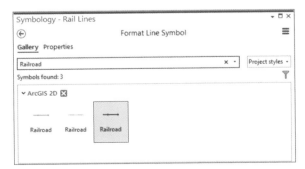

15. On the Edit tab, in the Selection group, click the Clear (Selected Features) button to see the new symbology.

16. In the Manage Edits group, click Save. Then answer yes to the Save Edits prompt to save and end your edits.

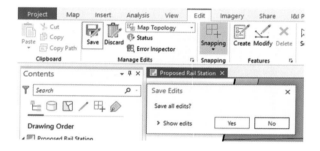

17. Save the project.

    This section concludes the first part of feature creation. You were able to take the rough schematic of the rail line location and, using the context menu and Feature Construction tools, draw the line in a precise location.

18. If you are not continuing to the next exercise, exit ArcGIS Pro.

Exercise 5-1

The tutorial demonstrated many of the line creation techniques that are available with the regular Feature Construction tools.

In this exercise, you will repeat the process for the second part of the diagram, which shows the extension of the rail line to the eastern end of the city.
- Start ArcGIS Pro, and open Tutorial 5-1.aprx, if necessary.
- Zoom to the Second Rail Segment bookmark.
- Snap to the end of the existing rail line, and extend it using these measurements:
  - An angle of **12.92** degrees (polar) for **375** feet
  - Parallel to the edge of the existing street for **4200** feet

- Save the results with your ArcGIS Pro project.

WHAT TO TURN IN

If you are working in a classroom setting with an instructor, you may be required to submit the maps you created in tutorial 5-1.
- Screenshot images of:
  - Tutorial 5-1
  - Exercise 5-1

Review

A successful geodatabase will contain accurate representations of important spatial features (i.e., points, lines, and polygons), important characteristics about these features in the form of attributes contained in an associated data table, and information about behavioral aspects of the relationships among features stored as a topology, as well as built-in rules that enhance the security and integrity of the data. There are several ways to populate these spatial features in the geodatabase. One way is to import existing features from data contained in shapefiles. As discussed, the ability to import existing data has tremendous opportunity for cost savings for an organization. However, there will be opportunities where you need to create new data as additional needs are identified in the organization. In this tutorial, the City of Oleander was interested in building a new light-rail system and had already identified an area within the city for this construction to occur. Existing data of the proposed rail system was not available in a digital format, so it was necessary to create this feature within your geodatabase. By using available tools in ArcGIS Pro, you were able to create this digital representation of the rail line from a paper map that had exacting specifications.

To ensure that your data editing is organized, it is best to organize how necessary features in the ArcGIS Pro session are displayed on the screen. In this tutorial, you created the proposed rail line feature based on a known starting point, and then used existing spatial features contained in other layers to complete the feature edits. As the accuracy of the proposed rail line depended on its relative proximity to other features, it was important that these features were drawn using the appropriate symbology and that they were organized and appropriately named within the Contents pane, so that important features were recognizable, accessible, and visible to the user. You created group layers that combined associated data layers and learned to customize how and when these features are drawn. You also examined how using bookmarks can help define areas of the map that you will need to refer to often. Lastly, by setting the default map extent, you saved valuable time and lessened the potential for getting "lost" by reducing the map extent from an "outer space" view to the project area. By following the steps in setting up the editing parameters within your project, you will find that you can work more quickly and efficiently.

Before you can edit or create any new data, you must have a container to store these features. The containers that store the data are the feature classes, which may or may not exist within feature datasets within your geodatabase. As you have already examined how to create feature classes and feature datasets in previous tutorials on developing the geodatabase, you should be familiar with the operations necessary to create them. In this tutorial, the geodatabase was provided, including the feature class to contain the proposed rail lines. This line feature class already contained the necessary spatial and attribute information that you needed to place your edits. However, for ArcGIS Pro to know in which feature class to place your edits, you had to explicitly select it from the Create Features template. Once you selected the appropriate feature to edit, and the appropriate construction tool, you could begin editing.

The number of tools available for creating and editing spatial features within ArcGIS Pro may seem a bit overwhelming at this point, but the more you use these tools in a variety of circumstances, the greater your knowledge of the capabilities of each one will become, and the more adept you will become at using them. The main thing to remember is, do not be afraid of testing each tool. That's why there is an Undo button on the toolbar!

STUDY QUESTIONS

1. List the different ways in which you can optimize your map as described in the tutorial. Which method is the most beneficial to you, and why?
2. What is a sketch in ArcGIS Pro? How is it different from a permanent feature?
3. List and define five different Feature Construction tools.
4. Illustrate an example of creating a feature employing at least three of the Feature Construction tools.

Other study topics

Search for these key phrases in ArcGIS Pro Help for further reading:
1. A quick tour of editing
2. Get started editing
3. Customize the editing tool gallery
4. Copy line features parallel

Tutorial 5-2: Using context menu creation tools

The first tutorial in this chapter barely scratched the surface of what can be done to create new features. This tutorial will look at more Feature Construction tools, as well as methods for using the selection and snapping controls to your advantage.

LEARNING OBJECTIVES
- Set snapping
- Set selectable layers
- Use the Feature Construction tools
- Create new features
- Use the context menu tools

Introduction

There are many more drawing tools available for creating new features than demonstrated in the last tutorial. Although tutorial 5-1 worked mainly with the context menu, these next techniques will use existing features and more sophisticated measurements to create new features.

One new concept for this tutorial is using the snapping environment. It will give you the ability to use existing features to help draw new features and to set, with some accuracy, which parts of existing features to use.

You can set each data layer to snap to the endpoints, vertices, or along the edge of a feature. An endpoint is where a feature begins and ends. For a line, this point is at each end, with the first endpoint shown as a green box and the last endpoint shown as a red box in the sketch. For a polygon, this is the single point at which the polygon begins, because a polygon must start and stop at the same point. This point is shown as a red square.

A vertex is a point along a line or the perimeter of a polygon at which the line changes direction but does not end. These vertices are shown as green squares in the sketch.

Points are easy, because you either snap to the location of the point, or you don't. Setting the snapping of a point to either an endpoint or a vertex is considered the same thing, and your new feature will snap to the exact x,y coordinate of the point.

An edge is any location along the line or polygon perimeter. The edge will let you snap anywhere along the length of a line.

In the GIS world, the importance of snapping features cannot be overemphasized. Many of the analysis tools used in GIS analysis require that the features have some connectivity, such as linear referencing or network analysis. If the features are not snapped, no analysis across the line features can be done. A snapping tolerance can be set, which represents the distance the cursor must be from the feature for the feature to snap to an existing location. The sketch also has specialized snapping settings.

Another new concept for this tutorial is setting the selection environment. When a map contains several layers, making a selection can be troublesome if unwanted features continually get selected. The solution is to implicitly set the selection environment. It is also possible to set the selection tolerance, or the distance in which ArcGIS Pro searches for features.

With these new settings, it is possible to use existing features to create new ones.

Using shortcuts to create features

Scenario

Certain features of the ORTA diagram are still not entered in the GIS feature datasets. Additional linear features relating to property boundaries and parking lots must still be drawn. You'll reference their diagrams and use more of the ArcGIS Pro Feature Construction tools to enter these next features.

Data

You will be using the same data as tutorial 5-1. The new features to enter will be the right-of-way along the railroad track and the parking lot area. Remember that creating features is a puzzle with the goal to discover which drawing tools can perform the desired tasks. Note also that in most circumstances, one tool won't draw everything you need. You may need to use a combination of tools and construction lines to draw the final features.

Tools used

ArcGIS Pro:
- Select Features
- Trace tool
- Clear Selected Features
- Bookmarks
- Snap to features
- Line: Parallel
- Line: Length

- Line: Perpendicular
- Trim tool
- Copy Parallel tool
- Split Feature tool

The first feature to draw will be the right-of-way for the railroad referenced in the diagram in tutorial 5-1. ORTA wants to purchase 27 meters of right-of-way on either side of the track, and you must show where that will fall. Since you'll be dealing with only a few of the datasets, you should set the selections to include only the Rail Lines and the Buildings and Paving Lines layers.

Set the selections

1. Start ArcGIS Pro, and open the project file Tutorial 5-2.aprx.

2. At the top of the Contents pane, click the List By Selection tab.

   Next to each layer is a check box that will allow the selectability of the layer, and a numeric display indicating how many features in the layer are currently selected and in a layer group. This example shows many features selected in the Property Base Group Layer.

   It is also possible to clear the selections of either an individual layer, or all the layers. By highlighting a layer, going to the > Selection group on the Map tab, and clicking Clear, only the selection for that layer is cleared. But by going to the Selection group on the Edit tab > and clicking Clear, all the selected features in all the layers are cleared.

3. Click the buttons to make all the layers not selectable, except for Rail Lines – Segment 1 and Buildings and Paving - Lines.

   Now in even the most congested areas of the map, if you click to select a feature, all the others will be ignored except these two. As a side note, if you press and hold the Ctrl key and click any selected layer, all the layers will be changed to not selectable. The reverse is also true—pressing and holding the Ctrl key and selecting an unmarked check box will make all the layers selectable. In fact, the Ctrl+Click technique works anywhere that ArcGIS Pro presents a list of check boxes, whether it's selections, layer visibility, activating field names, or other functions. This technique also works with the toggle key in the layers list to show/hide the legend of layers.

4. Try out the selection by going to the Edit tab and then the Selection group. Click Select, and click the railroad track drawn in the last tutorial. Try clicking an area where the rail line crosses other features. Note that only the rail line is selected. After trying the tool in other areas, clear the selected features before continuing.

5. Return to the List By Drawing Order tab in the Contents pane.

Edit and trace features

According to the diagram, the right-of-way (ROW) exactly parallels the proposed rail line by 27 meters on either side. It would be hard, if not impossible, to modify the provided measurement information for the rail line to make these ROW lines manually, but with the Trace tool, they can be traced with an offset to provide the desired results.

1. Go to the Edit tab, select the Features group, and click Create. The Create Features template will be added to your display.

2. In the Create Features template, click Proposed ROW. In the Construction Tools box, click Trace.

Remember to confirm the template feature selection in the Create Features pane each time you draw new features. Confirming this selection will keep you from drawing in the wrong layer.

This tool will exactly trace any feature. It will also jump between features if they are snapped and will even duplicate arcs as true curves. In this instance, you want to trace the existing line representing the railroad track but move it over 27 meters. You'll do the tracing using the Offset tool and an on-the-fly unit conversion.

3. Click the railroad, press *O* to open the Offset dialog box, and click the check box for "Trace with offset." In the Offset distance box, type 27 m, and press Tab. Because "m" is the abbreviation for meters, the distance units will automatically change to meters. Click OK.

The tracing will begin at the west endpoint of the rail line and end at the east end. It will be important to snap to the end of the line, so you will need to review the snapping environment.

4. In the Contents pane, click the List By Snapping tab. Clear the check box for each layer except for Rail Lines – Segment 1. Hint: You could also right-click the layer and click "Make this the only snappable layer" in the context menu.

5. Because none of the operations you will be doing will require an edge snap, you can disable that function. On the Edit tab, in the Snapping group, click the arrow under Snapping. In the pop-up box, clear the check box for each of the icons except End Snapping and Vertex Snapping.

Once the snapping is set, the snap tip in edit mode will indicate which feature type you are snapping to.

The Snapping menu also contains four specialty snapping settings: the Intersection Snapping tool will create a snap point where two features cross, even if there is no node at the location or if the features are in different layers. The Midpoint Snapping tool will snap to the midpoint of any line segment. The Tangent Snapping tool allows you to snap a line to be tangent to an existing curve feature.

6. Slowly move the cursor to the left end of the selected feature until the snap tip shows that you are snapping to the endpoint of the rail line. Click that endpoint to start the new feature. Now move the cursor slowly to the right, along the rail line feature. A new sketch is being drawn as an exact trace, with a 27-meter offset.

7. If you are having trouble getting the trace to follow the correct line, zoom in a little and be careful to stay over the rail line. Sometimes the Trace tool can go off on a different line from what you expect, but you can drag backward over the line to remove that part of the sketch and continue.

8. Move to the far end, watch for the snap tip to display Rail Lines: Endpoint, and click again. A sketch will be drawn with a red box indicating the last point entered.

There's another part to this ROW on the north side, but rather than having two different lines, you should make it a single, multipart feature.

9. Right-click to expose the context menu, and then click Finish Part.

   Notice that the rubber band from the end of the line is gone, but the feature is still in edit mode. Using the sketch means that you can draw more parts to the feature. The result will be a single record in the table, but there will be multiple graphic representations of that record. Being able to link multiple graphic items to a single record is useful for features that take multiple parts to represent their entire structure.

10. Move to the far right of the rail line until the cursor's snap tip displays Rail Line: Endpoint. Click to start the next part of the feature. Then move along the feature, again tracing the selected feature. When you reach the left end of the line, double-click. The cursor will snap to that end of the line and will end the feature creation.

11. **Clear the selected features to show the symbolized feature, and save your edits.**

Next, you'll add another feature to the feature class using some of the specialized Sketch construction tools. The trick to using these tools is to know them thoroughly and determine which ones can be used to achieve your goal.

You'll put in the perimeter of the parking lot. It is roughly dimensioned in the diagram and shown in relationship to the street and property line, but even then, some of the lines will be put in as a sketch that is not fully dimensioned.

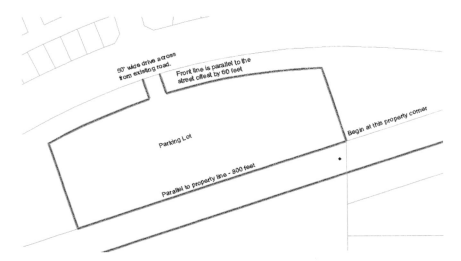

You'll start with the basic perimeter of the parking lot. The only point that is tied to any existing features is the southeast corner, which coincides with the property corner there. If you snap to that corner and go parallel to the ROW line, you will be able to draw most of the perimeter. Along the street, there are no lengths recorded, but it is noted that the front is 60 feet from the street edge and runs parallel to the curve. You can draw that portion as a separate feature and then tidy up the corners. The main entry is not located in reference to any other feature, but it is 50 feet wide and appears to line up with a street opposite the main boulevard. You can sketch the entry while keeping it the correct width and tying it to the existing street.

A lot of the context menu tools will be used here, and some of them, such as the parallel and perpendicular functions, can be invoked from the Feature Construction toolbar. Be careful with your cursor placement, especially when tracing, and make deliberate movements to help prevent errors.

Prepare for editing

A bookmark has been created for the extent of the new station's parking lot and station building. You can use this bookmark to zoom to the site.

1. On the Map tab, in the Navigate group, zoom to the Oleander Station bookmark.

2. Set the snapping layers to include Lot Boundaries, Building and Paving – Lines, and Rail Lines – Segment 1. Then set the snapping to allow edge snapping and intersection snapping.

Draw the parking lot

The parking lot will start at the property corner shown in the diagram under "Edit and trace features." The front dimension is given but not the sides. However, because they are shown to stop 60 feet from the edge of the existing road, you'll draw them longer than necessary and clip them later.

1. In the Create Features template, click Paving and confirm that the construction tool is set to Line.

2. Move the cursor to the property corner shown in the diagram until the snap tip shows Lot Boundaries: Endpoint. Click it to start the line.

   The sketch is started, and the rubber banding is in place. According to the diagram, the pavement runs parallel to the ROW line for 800 feet.

3. Move the cursor over the existing ROW line, right-click, and click Parallel to constrain the drawing angle. Next, right-click and click Distance. Enter a value of 800 (ft). This action will complete the first segment.

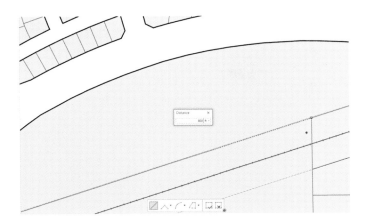

4. Move the cursor over the same ROW edge, right-click, and click Perpendicular. Move the cursor up near the street edge, and click to add a vertex. Don't worry too much about placement, since this will be trimmed later.

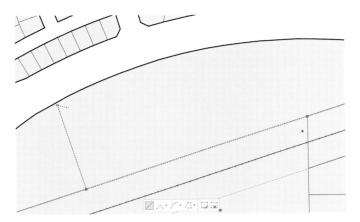

As you recall, the red box over the last vertex indicates that it was the last one entered. And new vertices will be added after this one. What needs to be done, however, is to add a line from the starting point that runs perpendicular to the ROW line. If you reverse the sketch, making the first point entered become the last point entered, the new line will continue from there. Confused? Try the command, and maybe the visualization will help.

5. Move the cursor to the sketch line you just drew, right-click, and then click Reverse Direction.

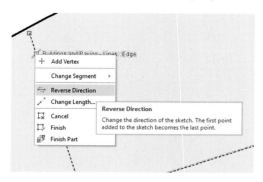

Now the red square is at the other end of the sketch, and new vertices will extend from this point. Next, you can use the Perpendicular command again and draw a line out to the street.

6. Use the Perpendicular tool and the same edge as before to constrain the new segment's angle. Move the cursor up near the street edge, and double-click to end the sketch. This segment will also be trimmed later.

The north edge of the parking lot runs parallel to, and 60 feet offset from, the existing street edge. Use the Trace tool, and draw a new feature that goes a little past the one you just drew. Then the overlaps can be cleaned up.

7. Select the Trace tool and move onto the street ROW until the snap tip identifies the correct line, and begin the trace. Set the offset distance to 60 (ft). If your tool starts tracing on the wrong side of the line, press the Tab key to switch sides. When you've completed the trace, double-click to create the feature.

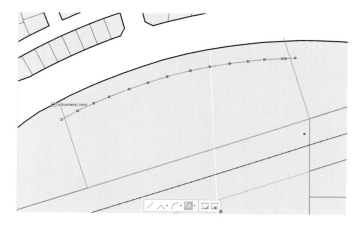

Chapter 5 | Working with features   163

You'll notice that the lines overlap. To clean up these lines, you will modify the vertices of the lines to move to the intersection point. This action will be aided by using the Intersection Snapping tool you activated.

8. **On the Edit tab, in the Features group, click Modify. This step will open the Modify Features pane. Under Reshape, click Vertices.**

9. **Click on the starting vertex and drag it toward the parking lot edge line until the snap tip indicates that you are at the intersection. Repeat with the opposite endpoint, and press F2 or Finish to complete the line.**

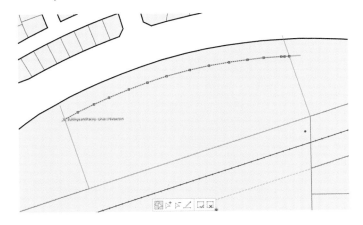

Both ends of the line are now trimmed at the parking lot edge. Next, you want to trim those lines to stop at the upper curved line.

10. In the Modify Features pane, click "Change the selection," and select the parking lot perimeter line. Move the endpoints to snap to the curve, and press F2 or Finish to complete the changes. Close the Modify pane when you are finished.

Clean up the line work

The perimeter of the parking lot is now drawn. Next, add the entry from the main road. The diagram doesn't show an exact placement, other than it lines up with the drive across the street. It's 50 feet wide, so you'll draw the first line perpendicular to the parking lot edge and trace it with a 50-foot offset.

1. On the Edit toolbar, click Create and select the line creation tool from the feature templates.

    You've done all these processes before, so the instructions here will be an overview from which you must determine the tools and placements to create the correct features.

2. Start your line at the corner of the ROW across from the station. Constrain it to be perpendicular to the edge of the parking lot, and draw it a little past the lot's edge. Draw a second line parallel to the first and offset by 50 (ft), again overlapping the other lines a bit. Finish the lines, and save your edits.

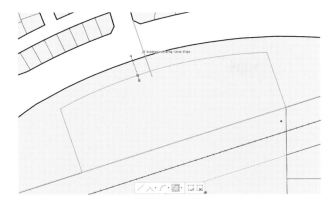

The lines overlap and can be fixed in one of several ways. You can edit the vertices and move them to the intersection points as before, or you can use the Trim tool and cut off the overlaps. But for this process, you will split the lines at the intersections and delete the remaining segments.

3. On the Edit tab, click Modify, and under Divide, select the Split tool. Select a line feature, if the last feature is not still selected, and move the cursor to the intersection of the two lines. Click to split the line.

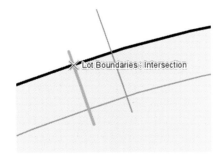

4. Repeat for the other intersections. You can also split the parking lot edge at the intersections, although you will not be able to split the lot line or ownership data. Select and delete all the line segments that were the overlaps, leaving the correct parking lot outline.

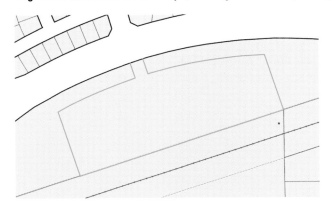

This step completes the creation of the parking lot perimeter.

5. Save your edits and the project file.

   You can see that the context menu and other drawing tools in ArcGIS Pro can be a little tricky, and it takes study and practice to determine which ones to use, and where. There are many more tools available, so it is advisable to examine the list of editing commands in ArcGIS Pro Help for more ideas on what the tools are and how to use them.

6. **If you are not continuing to the next exercise, exit ArcGIS Pro.**

Exercise 5-2

The tutorial showed more of the editing tools from the Modify Features pane.
In this exercise, you will repeat the process for a proposed rail station.
- Start ArcGIS Pro, and open Tutorial 5-2.aprx if necessary.
- Zoom to the Second Rail Segment bookmark, and turn on the Rail Lines – Segment 2 layer.
- Add the proposed ROW line for the extension of the rail line.
- Zoom to the Commuter Rail Stop bookmark.
- Use the dimensions on the layout diagram to add the parking lot for the Commuter Rail Stop.

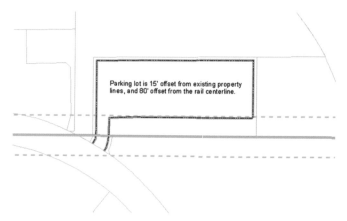

- Save the results when completed.

WHAT TO TURN IN

If you are working in a classroom setting with an instructor, you may be required to submit the maps you created in tutorial 5-2.
- Screenshot images of:
  - Tutorial 5-2
  - Exercise 5-2

Review

Creating spatial data in the geodatabase can be a time-consuming and complicated task. ArcGIS Pro provides a multitude of tools that, when learned and applied properly, will help tremendously.

Tutorial 5-2 continued the creation of the Oleander proposed rail station features in the geodatabase by illustrating the use of several standard and context menu tools to create a right-of-way around the proposed railway and create a parking lot for the terminal station. Before beginning your edits, you had to do two things that aided the editing process in ArcGIS Pro: set the selectable layers and establish the snapping environment. For a feature to be edited, it must be selected first. Therefore, you must make sure that the feature you are editing is selectable. The easiest way to do that is to click the List By Selection tab at the top of the Contents pane of your project and review which features in the list are selected via a marked check box. Not only does this marking let ArcGIS Pro know which features should be selected, but it also prevents any other features from inadvertently being selected for further action.

In the GIS world, the importance of snapping features cannot be overemphasized. Many of the analysis tools, such as linear referencing and network analysis, require that the features maintain connectivity. If the features are not snapped, no analysis across the line features can be done. The snapping environment provides the "rules of engagement" that will apply throughout editing or until otherwise changed for the features you are editing. Components of the lines, polygons, and point features that you create in the geodatabase are referred to as *vertices*, *edges*, and *endpoints*. Vertices are much the same as point features, except that vertices are connected by segments and make up line or polygon features. Point features and vertices are created using the same methods. Edges are any location along the line or polygon perimeter. Endpoints are the points at the end of a line segment.

Although the snapping environment is typically established at the start of editing, it can be established and/or modified at any time during editing. In addition, these rules can be intentionally bypassed by using the right-click context menu for one-click snapping when editing a feature.

Many times, your project may call for specific editing tasks to be used more than others. Task options can be added or removed from the different toolbars and menus. This organization can simplify your editing interface.

STUDY QUESTIONS

1. List and define five tools from the Modify Features pane.
2. List and define the three different features set in the snapping environment. Why is it important to specify the snapping behavior during editing?
3. Why is it important to define a selectable layer? Explain two different ways of making a layer selectable in ArcGIS Pro.
4. Is there another way to create a right-of-way delineation of 27 meters on either side of the proposed rail line? Explain your answer. Would you arrive at the same result? How?

Other study topics

Search for these key phrases in ArcGIS Pro Help for further reading:

1. Introduction to modifying features
2. Modify Features tool reference
3. Create segments by tracing
4. Keyboard shortcuts for editing

Chapter 6

Advanced editing

Often, the process of drawing new features into an ArcGIS Pro project requires a lot of problem solving. You are given some information describing spatial data, and you must figure out which tool would be most suitable for entering the data. These tools can get complex, or you might need a combination of tools to complete the task, so knowing the characteristics of the full range of ArcGIS Pro editing tools is essential.

Tutorial 6-1: Exploring more creation tools

The Feature Construction tools worked well in the past two tutorials in chapter 5 to create new linear features, but how well do they work on polygons? This section will deal with creating polygon features using the context menu functions and a few choices that are specific to polygons.

LEARNING OBJECTIVES
- Use Feature Construction tools
- Create new features
- Use the context menu tools
- Work with polygons

Introduction

All the features used in GIS must be represented by points, lines, or polygons, so it is important that there be good, strong tools for making polygons, just as there are for creating lines. Polygons have the unique characteristic that they must start and end at the same place, so locating the starting point means that you have also located the endpoint. But for all the other vertices in a polygon, you must rely on the regular ArcGIS Pro construction tools. And just as with the lines, the ability to solve the puzzle of which tools to use when will make you successful in using them.

This tutorial will look at more of the Feature Construction tools and context menu functions as they relate to polygons, but the same ideas can be applicable to points and lines as well.

More tools and functions

Scenario

In the ORTA diagram for the new North Oleander Station, you can see that the buildings need to be added. They are represented by polygons, and some information about their locations and dimensions is given. Using the ArcGIS Pro tools and context menu functions, you'll add the rest of the features to the drawing.

The first building to add, the transit station, will be the large one that parallels the tracks. It can then be used as a reference point to add the smaller buildings.

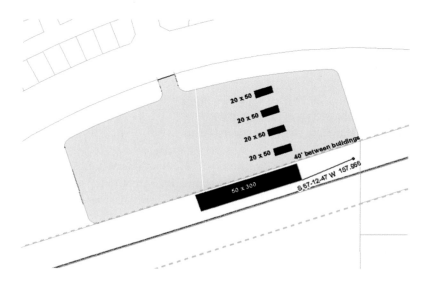

Data

You will be using the same data from tutorial 5-1.

Tools used

ArcGIS Pro:
- Line: Direction/Distance
- Line: Delete Vertex
- Line: Perpendicular
- Line: Distance
- Line: Parallel
- Line: Square and Finish
- Copy/Paste
- Trace tool
- Symbol levels

First, you will need to take care of some setup procedures such as setting the selection, editing, and snapping environments.

Setting up editing

A lot of snapping will be going on in this tutorial, so you should set the snapping environment to include all the features you set in tutorial 5-2.

1. **Start ArcGIS Pro, and open the project file Tutorial 6-1.aprx.**

   You will see the familiar layout of the proposed Oleander rail system.

2. **In the Contents pane, use the different tabs to set these constraints:**
   - **Only the Building and Paving – Polygons layer is selectable.**
   - **Only the Building and Paving – Polygons layer can be edited.**
   - **Snapping is allowed only for the Building and Paving – Polygons layer, the Survey Control Points layer, and the Building and Paving – Lines layer.**
   - **Snapping is on for Point, End, and Vertex snapping (no edges or intersections).**

   If you are unfamiliar with how to set these parameters, review tutorial 5-2.
   Next, you'll need to set the units for the direction measurements. The default is polar coordinates, where zero degrees is on the right and angles are measured in degrees counterclockwise. The information provided is in quadrant bearing, a measuring system used by surveyors. The angle describes which quadrant the angle is measured in, with a measurement in degrees/minutes/seconds.

3. **On the Project tab, click Options. Under Current Settings, click Units and then the Direction Units tab. Set it to Quadrant Bearing. Click OK, and then click the back arrow to return to the project.**

The units may have been set in a previous project and may already be the value that you want, but it is important to check and verify the setting before beginning your edits.

Draw the buildings

With all the necessary parameters for the data entry environment set up, you are ready to start drawing the buildings. The corner of the first building is referenced to the survey control point in survey measurements. You could draw a temporary line to represent the distance, but with some cleverness, it can be done without it.

1. Zoom to the bookmark North Oleander Station.

2. In the Create Features template, select the Buildings and Paving - Polygons layer with the standard Polygon tool selected.

3. Move the cursor to the survey control point until the snap tip displays Survey Control Points: Point. Then click to start a sketch.

4. Invoke the Direction/Distance tool from the context menu, and enter the dimensions as shown in the diagram at the start of this tutorial:
   S 67-12-47 W, 157.966

5. Press Enter to store the coordinates and create the vertex in the sketch.

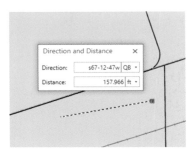

6. Move the cursor back over the starting point of the sketch, right-click, and then click Delete Vertex.

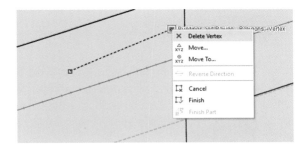

This step has used the survey control point and survey measurement to locate the starting corner of the building but has removed the vertex so that it won't be included in the final feature.

The front and back of the building run parallel to the track, with the sides perpendicular to the tracks and measuring 50 × 300 feet. Using the Feature Construction tools, you should be able to draw the building.

7. Set the sketch constrained to parallel to the rail line and make its length 300 (ft), and then press Enter. Next, constrain the line so it's perpendicular to the rail line, make its length 50 (ft), and press Enter. Note: The Parallel tool may mimic the direction of the line you use as a reference, so you may have to provide a length of −300 (ft) (opposite the direction of the original line).

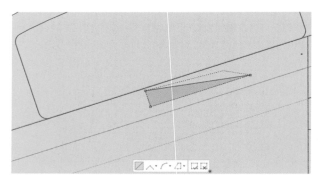

These steps made half the building. Now it's time for a special tool that can be used with rectangular polygons.

8. **Right-click, and click Square and Finish on the context menu.**

This tool will square off the building and create the feature. It only works with polygons because ArcGIS Pro already knows what the last vertex is going to be—it's the same as the starting point.

YOUR TURN
After consulting the layout diagram, you will see that the next building aligns with the east end of this building and is offset by 40 feet. It appears to run parallel/perpendicular from this building as well. Use the same technique to draw the new 20 × 50 building. Start at the known corner, draw a line of the specified direction and distance, remove the original vertex, draw two sides of the building, and use Square and Finish to complete it.

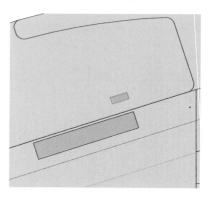

The next buildings are repeats of the 20 × 50 buildings, also set 40 feet apart. Rather than draw them, you can copy one and move the copy into the correct position. The trick, however, is to place it with precision.

9. Make sure the last feature drawn is selected, and then go to the Edit tab. In the Clipboard group, click Copy and then Paste. The feature will be pasted back into the currently selected template. Notice that the Modify Features pane automatically opened when you pasted the new feature.

An exact copy of the building has been added to the map, but it is directly on top of the first building. You will need to move it but in reference to the other building. The yellow dot shown in the middle of the selected feature is its attachment point and can be used to snap to other features. The problem is that the dot is in the wrong place, but it can be moved.

10. In the Modify Features pane, select the Move tool.

11. Move the cursor over the yellow dot, and then press and hold the Ctrl key. The cursor should turn into a little box with arrows around it. Drag the box down until it snaps to the corner of the original station building, and then release the Ctrl key.

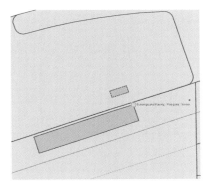

12. The cursor will now become a four-headed arrow again. Drag the feature away from the building you copied. Notice that the yellow dot stays in relative position to the copy of the feature. Move up until the dot snaps to the upper-right corner of the second building, and then release the mouse. The building is now placed in its correct position, which is 40 feet away from the other building. Click F2 or Finish.

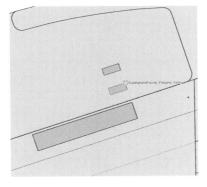

This maneuver can be tricky, but if done correctly, it can be a great time saver. Practice this step until you can do it smoothly.

The copy/paste/move technique would work for adding two more buildings, but in the interest of exploration you will add the next two using the edit grid. The edit grid is a temporary grid of guidelines that can be superimposed on your drawing and used to add new features. The grid can be made any size and can be snapped and rotated to existing features. Instead of always using the existing building to find the location of another building, you will set up a grid to determine those spacings.

Set up an edit grid

1. **On the bottom of the map pane is a set of icons that control snapping, dynamic constraints, edit grids, and COGO conversions. Hover over the third icon to display the grid controls.**

2. **Set the grid size to 10 (ft), and then click the Set Origin and Rotation button.**

Two things happen when you click the Set Origin and Rotation button. First, the grid of the specified size is displayed on the map. Second, the grid toolbar is displayed at the bottom of the map pane. This toolbar has a tool to set both the origin and rotation, a tool to set just the origin without changing the rotation, and a tool to set the rotation without changing the origin.

3. **Move the red cursor over to the lower left of the newest building that you just drew, and click to snap to the corner. Then move the cursor over to the other end of the building, and snap to the corner to align the grid with the building.**

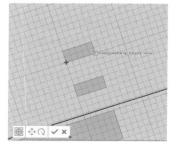

4. Use the Create Features polygon tool to draw a new building by counting up four grid lines from the corner of the building and drawing a box 20 × 50 feet. Make sure each corner snaps to the grid.

5. Repeat for the last building. Click the Grid button to turn the grid off.

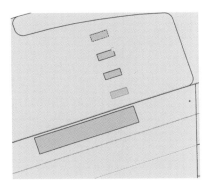

Symbolize unique features

The city planner wants to know how many square feet the parking lot is and wants to symbolize it differently on the map.

Drawing the feature into the Buildings and Paving - Polygons feature class will be an easy trace task. But there should be a unique way to identify the parking lot so that it can be symbolized differently. The attribute table has a field named Description that will help. The first building is the transit station, and the outbuildings are bus shelters. If these buildings are added, and the parking lot is given a unique description, it can be symbolized that way.

1. In the Contents pane, right-click the Buildings and Paving - Polygons layer, point to Selection, and click Select All. Next, on the Edit tab, in the Selection group, click the Attributes button to open the Attributes pane.

Five features are shown. One is the transit station, and the others are bus shelters. A fast way to populate the attributes is to populate them globally to Bus Shelter, and then discover which one is the station and change it.

2. At the top of the dialog box, click Buildings and Paving - Polygons. When this top row is selected, any changes you make will be applied to all selected features. Next, click in the empty line next to Description. Type Bus Shelter, and click Apply.

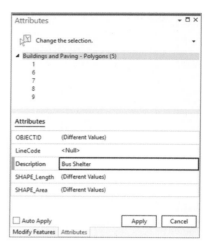

3. Click each of the features, and watch in the drawing until the station feature flashes. When you find it, change its description to Station, and click Apply. Close the Attributes pane, and save your edits.

   With the correct description taken care of, you can draw the polygon for the parking lot. But don't forget to set its attribute value. You can set an option that will automatically populate the attribute information each time a feature is drawn.

4. In the Create Features template, select the Buildings and Paving - Polygons layer, and then click the arrow to open Properties. On the blank line next to Description, type Parking Lot, and then click the arrow to return to the Create Features pane.

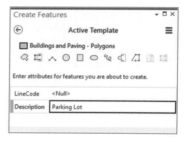

   This value will now be the default value for all new features entered. You're only drawing one, but if you were drawing many more parking lots, this default value would be a time saver. Always remember, however, to change the default value before drawing other feature types.

5. Select the Trace tool, make sure the Buildings and Paving - Polygons layer is selected in the Create Features template, and trace the perimeter of the lot. When the entire lot is traced, double-click to finish the feature.

Chapter 6 | Advanced editing   179

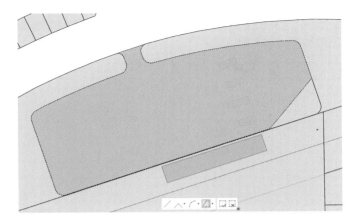

The feature was drawn correctly, but now it obscures the bus shelters. To fix this problem, you will change the symbology of the feature.

6. **In the Contents pane, select the Buildings and Paving - Polygons layer. On the Appearance tab, in the Drawing group, click the drop-down arrow under Symbology and click Unique Values.**

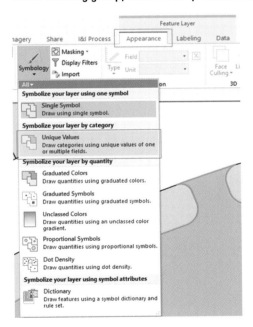

7. **Set the Value field to Description. Click the More button, and clear the "Show all other values" box.**

8. Change the colors to Electron Gold for the bus shelters, Gray 20% for the parking lot, and Ultra Blue for the station. Examine the results in the map.

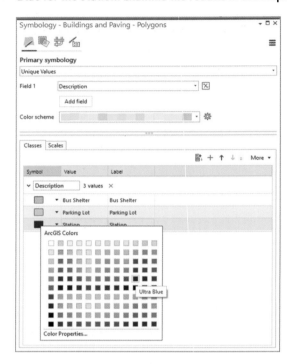

Well, that symbology almost worked. The parking lot did, in fact, draw in the chosen color, but now it obscures the bus shelters. There is a way to fix this problem by using an advanced setting for symbology.

9. In the Symbology pane, go to the Symbol layer drawing tab. Enable the symbol layer drawing using the slide switch, and drag the parking lot to the bottom of the list.

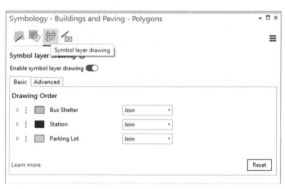

This maneuver seems to have done the trick. The parking lot drawing is at the bottom of the layers, and the bus shelters are now visible. Symbol Layer Drawing is a useful feature to control the display order of features within a single feature class. Previously, the features would have been drawn in order of their ObjectID number, but now that control has been turned over to the user.

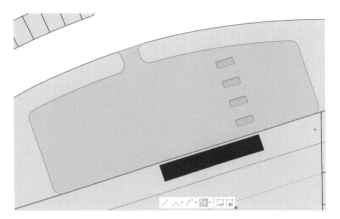

10. **Save your edits, and save the project file.**

This tutorial concludes the work with the ORTA diagrams. This exhibit can now be shown to the City Council for reference and used to select the number of households within a specified distance of the station or to examine the utilities in the area to determine whether there is any work the city may have to complete.

If you are not continuing to the next exercise, exit ArcGIS Pro.

Exercise 6-1

The tutorial showed how to use the same context menu tools to draw polygons as you do for points and lines. It also demonstrated the Symbol Layer Drawing controls.

In this exercise, you will repeat the process for the buildings and parking lot at the commuter station.
- Start ArcGIS Pro, and open Tutorial 6-1.aprx, if necessary.
- Zoom to the bookmark Commuter Rail Stop.
- Use the trace technique to draw the parking lot.
- Draw the buildings as shown with the direction/distance value to the starting corner of:
   N 31 deg 58 min 42 sec W, 1110.819 feet

- Populate the attribute table with the correct descriptions.

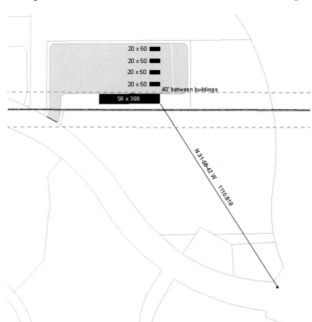

- Save the results when completed.

WHAT TO TURN IN

If you are working in a classroom setting with an instructor, you may be required to submit the maps you created in tutorial 6-1.
- Screenshot images of:
  - Tutorial 6-1
  - Exercise 6-1

Review

Creating polygons in ArcGIS Pro is like creating lines. With polygons, the difference is that the feature you create will close at the same location where it began. All the feature editing thus far has relied on the same initial steps: adding the feature class to your map, selecting the feature class for editing, establishing your snapping environment, specifying the feature to draw and the tool to use in the Create Features template, and beginning the edit process.

As you could see, the same tools, context menus, and symbology available when creating line features are available when creating polygons. Creating polygons was a good opportunity to continue to practice and learn these tools for later use.

Because you were relying on a survey document to create your station and associated buildings, you had to set the distance measurements for your sketch polygon that were consistent with the measurements contained in the survey. This measurement was necessary so that you could accurately place the building in the proper location, as well as construct the shape of the station as intended in the survey plan. Because you used the bus station as a relative location to place the associated bus shelters, it was critical that the location and dimensions of the bus station be as accurate as possible.

Having the ability to edit attributes of features in the feature template is a convenient way of attributing data as you create it. By selecting all the shelters, you were able to attribute multiple buildings simultaneously. This advance attribution not only prevented potential errors in editing the data, but it also prevented you from having to go in afterward and add the attribution.

A little common sense goes a long way when you are creating features in ArcGIS Pro. Think through your objectives and develop an editing plan to outline your objectives and examine what tools are available in ArcGIS Pro. As you become more familiar with how all the tools work in ArcGIS Pro, you will be able to combine this knowledge with some ingenuity to figure out how to solve a puzzle in the most efficient manner.

STUDY QUESTIONS

1. In this exercise, you were given the distance between buildings to be created. You were able to use this information to determine the length and width of the buildings. How could you use the editing tools to create these buildings if you were given the coordinates at only one corner of each building?
2. Could you have used lines instead of polygons to represent the buildings? What would be an advantage and a disadvantage of using this approach?
3. Name and describe some of the context menu tools available for polygons that were not available for lines.

Other study topics

Search for these key phrases in ArcGIS Pro Help for further reading:
1. Edit feature attributes
2. Symbolize feature layers
3. Symbol layer drawing
4. Copy and paste using the clipboard

Tutorial 6-2: Building feature templates

When creating new features, it is common to use the default feature templates to draw them. They will go in as single lines, points, or polygons. Meanwhile, advanced templates can be customized to draw features as multiple items, and even divide them into multiple feature classes.

LEARNING OBJECTIVES
- Create new features
- Build group templates
- Create a preset template

Introduction

Anytime you start creating features, the standard Create Features pane will open. It will display a template for the features in each layer that are set as editable. The features will match the classifications on the layers, displaying the colors and shapes that have been set. Using these style templates, however, you can create only one feature at a time.

When multiple features must be drawn to represent a certain item, or features in different layers must be coincident, it may be advantageous to draw them with a group template. Group templates are constructed prior to editing and build a framework of features that can be created with a minimum amount of user interaction. In situations in which one feature type must always exist when another feature type is drawn, the group template can cause those features to be drawn automatically. For instance, if you are drawing the boundaries of swimming pools from aerial photography, you may want to create a polygon feature to represent the water of the pool; in addition, you may want a line feature around the perimeter of the pool to use for various types of symbolization. A group template can be built so that when you draw the polygon of the water, a linear feature will automatically be created in a different feature class coincident with the polygon edge. These custom templates are set up using feature builders.

Many types of feature builders are available, and a complete list can be found in ArcGIS Pro Help in the Feature Builders reference table. The user creates the primary geometry item (point, line, or polygon), and the builder can add other template geometry of any type and from any feature class. When a point is used as the primary geometry, the template can add additional points or a buffer around the point. A line drawn as the primary geometry can prompt the template to automatically draw other features of any geometry type, such as a point feature at the line vertices, a polygon buffer, or even additional lines coincident to the primary geometry. If a polygon is drawn as the primary geometry, any other geometry type can be automatically created. The templates can also be stacked to create complex features. For instance, the user might draw a polygon with a group template that adds perimeter line work, and that line work is a group template that adds points at every vertex. With some forethought, group templates can be used to enhance both the speed of feature creation and the data integrity.

The last type of template is the table template. This type of template is used to create a new row in a related table whenever a new feature is created in the primary feature class. For instance, creating a new ownership polygon feature can create a new record in the related tax roll database with predefined defaults. The template can also add records to multiple related tables, so the same ownership polygon could also create a new record in the related water service table, or more. Imagine a workflow in which a new ownership polygon is drawn, and then records are automatically created in the related tables to establish an ownership record, a water and electric service, a building permits account, and a membership with the Parks Department for use of the city's workout center.

The key to using these templates is determining which feature should be the primary geometry, and which geometries can be created on the basis of the new feature that is drawn. The most likely candidate for the primary geometry is a polygon feature because it can create any of the other feature types. Conversely, the only way to automatically generate a polygon feature from a point or line is with the use of buffering. However, if the new features do not contain a polygon feature, the primary geometry would most likely be a line. A line can generate points and other lines using many different styles. The last choice, a point, can duplicate only itself or create a buffer around itself providing limited options.

In addition to the "draw as you go" templates, another template, the preset template, uses existing features to make a "rubber stamp" template. Rather than having the user draw a primary geometry item and build features based on that, this option makes a template from a selected set of features and duplicates them exactly at any given location. For example, a set of electrical transformers may have the same grouping and position on every power pole, so a template could be made from a set of these existing features. Once created, the user could drop in the full set of transformers with one mouse click. These features can be easy to make but may lack the flexibility of the other template types since they will always be duplicates of the originals.

Building group templates

Scenario

The GIS manager for the City of Oleander has built several group templates for adding data. These templates include one for drawing new building footprints and their associated perimeter line, one for drawing street culverts and the headwalls at the ends, and one for adding a water system bypass with a control valve and a blow off valve.

In addition, the GIS manager would like to build three new templates. The first is for adding sewer lines and manholes. Every time a sewer line starts, stops, or makes a turn greater than 15 degrees, it must have a manhole. Because this condition is an absolute, he would like to help automate this process using a group template.

The second template is for adding a fire hydrant, the 6-inch PVC line that connects them into the water system, and the hydrant control valve. Again, this group of features is an absolute, so perhaps creating them can be automated.

The third is a set of features that make up a storm drain component for removing trash from the passing storm water. It has a point feature to represent the cyclonic chamber that sorts and traps trash, two point features to represent manholes for cleaning the system and removing the trash, and a polygon feature that represents the footprint of the concrete box that is installed with the unit.

You will start by creating features using the existing group templates and investigating their structure, and then you will create new group features for the sewer and water systems.

Data

Included in the map are the feature classes for the building footprints and outlines, along with the utility data for water, sewer, and storm drains. The layers will be used to create new features using group templates.

Tools used

ArcGIS Pro:
- Group template
- Feature builder
- Create Features
- Preset template

Using group templates

1. Start ArcGIS Pro, and open the project file Tutorial 6-2.aprx.

2. When the project opens, make sure the Group Templates Demos map is open, and zoom to the Building Footprints bookmark.

Chapter 6 | Advanced editing

The display shows the Landing subdivision in Oleander. The aerial photo shows that more houses were constructed in this subdivision since the last updates. They are represented with a yellow polygon that traces the roof line. In addition, they have a liner outline that duplicates the roof perimeter and includes an offset line that represents the edge of the constructed building.

These polygon features are created by tracing the buildings from the aerial photos, and since they mostly contain right angles, the right-angle tool helps speed their creation. First, you will draw a few of these house features using a group template, and then look at the properties of the template to see how it was created.

3. **Zoom to the House 1 bookmark. Turn off the Ownership Parcels layer to make the aerial image clearer.**

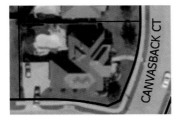

4. **On the Edit tab, in the Features group, click Create to open the Create Features pane. Within that pane, locate and click the Residential Footprint w/Outline template. Note the tool description, which shows that both polygon and line features will be drawn with this template.**

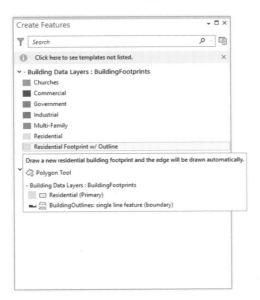

5. In the map, draw a polygon along the roofline of the house. Setting the Building Footprints layer transparency to 50% may help you see the buildings as you draw over them. After the first line segment is drawn, you may want to switch to the Right Angle Line construction tool to keep all the corners of the building square. It is also useful to use the Square and Finish context menu option to keep the last segment square as well.

6. When the feature is finished, clear the selected features to see the result.

The house is now represented with a yellow polygon, a dashed line, and a solid line—and all these features were created from the single drawing action you took.

Chapter 6 | Advanced editing 189

YOUR TURN

Twelve more houses were added on this block since the last update. Use bookmarks House 2 through House 13, and add the new house in each view.

When all the house outlines are added, clear the selected features and set the transparency back to 0% if necessary. Then go to the Edit tab, and save your edits.

The custom group template made drawing multiple features for each building easy, but how exactly does it do that? To discover the secret, you must look at the Manage Templates pane.

7. On the Edit tab, in the Features group, click the options launcher in the lower-right corner.

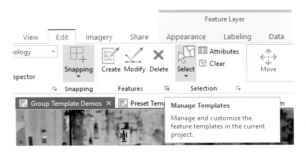

8. In the Manage Templates pane, select the Building Footprints layer, and then at the bottom, select the Residential Footprint w/Outline template.

9. With the template selected, right-click Properties to see the components of this template.

10. The General tab shows the name and description of the template and the target layer and symbol being used. Note that tags are added to allow for searches.

With the target layer set as BuildingFootprints, a polygon layer, the primary geometry will be polygons. The user will draw a polygon, and all other features will be created in relation to that new feature.

11. In template properties, under General, click the Tools tab.

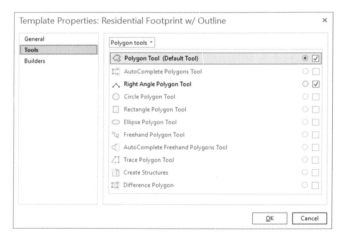

On the Tools tab, the template designer chose all the tools that the user will be allowed to access while creating features with this template. Any of the tools could create the building footprint polygon feature, but not all are appropriate for this task. Only the tools with their boxes checked will be activated for this template, and the tool with the radio button checked will be the default. For this template, only the Polygon tool and the Right Angle Polygon tool are active, and the Polygon tool is set as the default. By controlling available tools, you can keep the user from selecting something inappropriate.

12. **Click the Builders tab.**

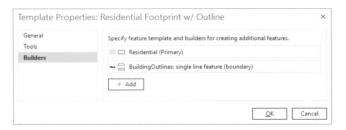

The dialog box shows the two features that will be constructed with this template. The first is the primary geometry from the BuildingFootprints layer, and the second is a line feature in the BuildingOutlines layer. Note that only a single outline feature is added, but it is symbolized using a double line. The Add button is used to create more features linked to the primary geometry, and that option will be explored more later in the tutorial, under "Creating simple group templates."

13. **Click OK to close the properties pane, and close the Manage Templates pane.**

More group templates

The next group template that you will look at is used to draw storm drain culverts—these culverts consist of a single pipe that runs under a road to carry storm water from one side to the other. Since these culverts take in water from a natural creek and expel the water into a natural creek, they have concrete headwalls at each end to prevent erosion. They are typically one straight segment of pipe with no bends or other fixtures attached to them except for the two headwalls. Since the pipe must have headwalls, the group template is set up to automatically add these features when the pipe is drawn. From this description, can you determine what the primary geometry will be?

Before drawing any culverts, you will first investigate the group template to see its components.

1. **Open the Manage Templates pane.**

2. **Select the Culverts layer. Then select the Storm Culvert with Headwalls template, right-click it, and click Properties.**

The General tab shows the name, description, and tags. Did you guess correctly on which feature would be the primary geometry? You can see that it is the linear feature Culverts.

3. **In Template Properties, click the Tools tab.**

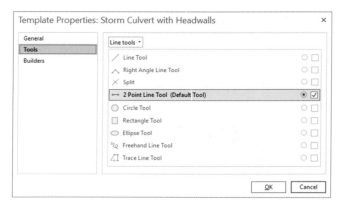

Culverts are always a single, straight pipe, which can be drawn with a line that has only the starting point and the endpoint with no internal vertices. It cannot contain curves or angles. Therefore, the 2 Point Line tool is the only one that will be presented to the user. This tool will not only simplify the input, but will also keep the user from misrepresenting the culverts.

4. **Click the Builders tab.**

Chapter 6 | Advanced editing   193

You can see that besides the line feature for the primary geometry, there is a feature builder that will add a point from the Headwalls layer at every vertex that is drawn. The line is locked into the 2 Point Line tool, which means that only two headwalls will be drawn—one at each end. That format matches the desired representation of the culvert.

5. **Close the Properties and Manage Templates panes.**

6. **Zoom to bookmark Culverts 1.**

   In the map display, you can see a road with a creek coming up to either side of it. There are culverts under the road, and the Oleander Public Works field crews have identified them by adding a label across them. You will draw in the culverts using the group template and covering each of the -C-U-L-V-E-R-T- labels on the map.

7. **On the Edit tab, in the Features group, click Create to open the feature construction templates.**

8. **Select the Storm Culvert with Headwalls template.**

9. **Click at the start of one of the -C-U-L-V-E-R-T- labels, as shown in the figure, and then click at the end of the label.**

A new feature for the culvert is drawn, along with two headwalls. Note that only two mouse clicks are required to draw the features.

10. Continue drawing the other two culverts that are part of this street crossing, making them parallel with the first one.

YOUR TURN

Move to the Culverts 2 bookmark and draw the culverts indicated for that street crossing. Repeat the process for the Culverts 3 and Culverts 4 bookmarks. Then turn off the Culvert Annotation layer to see the finished features.

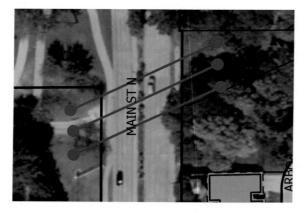

When you're finished, save your edits and the project.

Without the group template, the user would have had to click to start the line, double-click to end the line, click to change templates, click to add the first headwall, and click to add the second headwall—six mouse clicks reduced to two. In addition, there's no way that the segment can be anything other than the proper two-point line, there's no way to add more headwalls than needed, and it's assured that the headwalls will snap to the line endpoints regardless of the snapping environment. These rules all combine to speed up the process and ensure data integrity.

Using preset templates

The next template that the Oleander staff has is a preset template. These templates allow the user to drop in a set of multiple features with a single mouse click. Often referred to as the *rubber stamp template* because of the way it functions, it is created from a set of existing features that the user wants to duplicate in multiple locations.

1. Move to the Preset Template Demo map, and zoom to the Air Release Valve bookmark.

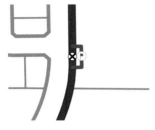

   This set of features represents a bypass gate valve system with a built-in air-release valve. It includes a point symbol for the gate valve, a blue line symbol for the bypass pipe, and another point symbol on the line for the air release valve. These bypasses function to isolate parts of the main and release any trapped air in the lines. Every time one of these bypasses is added, it includes this exact same set of features, making it a perfect candidate for a rubber stamp template.

2. Open the Manage Templates pane, select Air Release Valves, and open the properties for the ARV Bypass template.

3. Move between the tabs, and note the parameters that are set for this type of template.

   You will notice that this type of template has no primary geometry, even though it can create features in three different layers. That's because all the features are drawn at once, and none are used as the basis to create secondary features. Notice also that the only available tool is one that draws a single point.

4. Close the Properties and Manage Templates panes.

   When the data was initially created more than 20 years ago, these bypass valves were not drawn as part of the system because their complexity interfered with the utility routing program in use at the time. Now, however, the new utility program can not only use these bypass valves but also estimate the pressure at the air-release valves and predict when maintenance work is needed. To help add these features, the Oleander Public Works field crew has GPS'ed the location of the bypass systems along one of the new water mains. You will need to add a bypass system at each of the GPS locations. As the field crew does more inspections and GPS work, there will be more of these systems to add in the future.

5. Zoom to the ARV_GPS1 bookmark.

6. Open the Create Features pane, and select the ARV Bypass template.

7. Move the snap point over the GPS location indicated by the red dot, and click to add the bypass system.

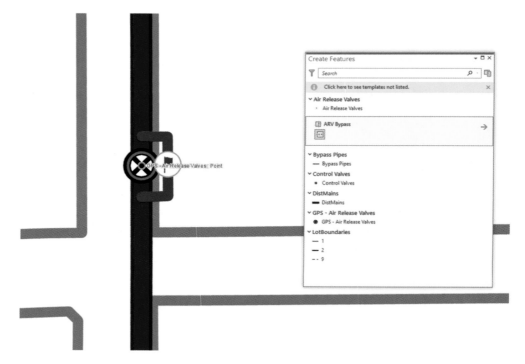

That's all there is to it. One click, and a complex set of features is added.

8. Add additional bypass systems at bookmarks ARV_GPS2 through ARV_GPS6.

9. When you're finished, close the Create Features pane, and save your edits.

The preset templates are an amazing tool for adding repetitive and complex systems with a single click. You will see more on how to create these templates later in this tutorial.

Creating simple group templates

Now that you've seen group templates in action and have examined some of their properties, you are ready to create new group templates.

The first template you will create is for adding sewer lines and manholes. Each time a sewer line changes direction or connects to another sewer line, it must have a manhole. These must-have scenarios usually make perfect opportunities for using a group template.

In this example, you will draw new sewer lines, one line segment at a time. These lines will typically start at an existing manhole and end at a location where a new manhole is needed. They can include several segments and even curves, so the choice of drawing tools will be broader than what was used for the culverts.

1. **Move to the Wastewater System map, and zoom to the Sewer Lines bookmark.**

2. **In the Contents pane, turn off the OwnershipParcels layer to make the sewer system more clearly visible.**

   This map includes a scanned, rectified image of the new wastewater lines that were installed with this subdivision. The new line locations are highlighted in yellow. The new group template will need to draw a line feature into the SewerLines layer and a point feature into the SewerNodes layer. Which should be the primary geometry? Could entering a single point location automatically determine the line location? No, there's not enough information from a single point to create a multipoint line. But entering a line location can indicate the location of a point feature—the single location at the end of the line. Then the primary geometry should be the sewer line, and a build rule can be added to create the manhole.

   Now that you know what the primary geometry should be, you can look at the build rules and determine which rule to use. The rules for creating a point with the primary geometry being a polyline are shown in the figure.

Primary geometry	Template geometry	Builder
Polyline	Point	Point at end of line
Polyline	Point	Point at start of line
Polyline	Point	Point at every vertex
Polyline	Point	Point at every vertex except start
Polyline	Point	Point at every vertex except end
Polyline	Point	Point at every vertex except start and end

   The editing plan is to start drawing new sewer line features at existing manholes and have the template automatically add a manhole at the end of the line. Thus, the "Point at end of line" builder is the one to use. Now you can start creating the new group template.

3. **Open the Manage Templates pane.**

4. **Select the SewerLines layer, and then click New > Group Template.**

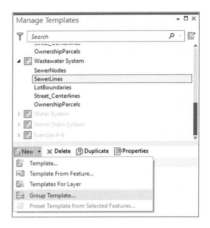

The Template Properties box will open, and this dialog box is used to create the new template.

5. **Enter the name** Sewer Line with Manhole.

6. **Provide this description:** The user draws a sewer line segment and the manhole is automatically added at the end of the line.

7. **Add a tag of** Oleander.

8. **Move to the Tools tab.**

   Look over the available tools and determine which tools you feel should be made available for the user.

9. **Leave the Line Tool active and marked as the default tool. Then clear the other checked tools that you have determined aren't appropriate. Your choices may differ from the figure.**

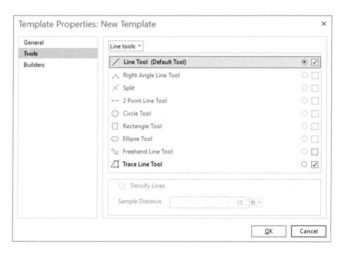

10. Click the Builders tab. The Primary Geometry item is shown.

11. Click Add, then click "Choose a feature template," and select manhole from the list.

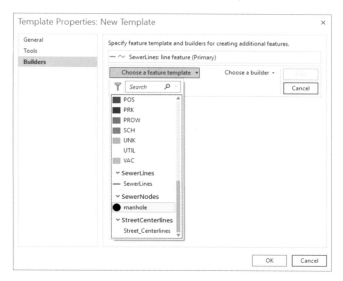

12. On the Choose a Builder button, select "point at end of line" and click Add.

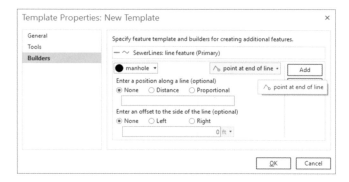

13. This addition completes the construction of the group template, so click OK to close the Properties pane. Then close the Manage Templates pane.

   Now you can start drawing new features. The process will be to snap to an existing manhole, draw the sewer line segment to the next manhole, and then double-click to finish. If things go as planned, a manhole will be drawn at the line's endpoint.

14. Zoom to the Sewer Segment 1 bookmark, and open the Create Features pane.

15. Select the Sewer Line with Manhole template. Click the existing manhole at the top of the display, draw a line to the manhole on Pintail Parkway, and double-click to finish.

16. Snap to the manhole you just created, draw the line to the manhole location in the image to the west, and double-click to finish.

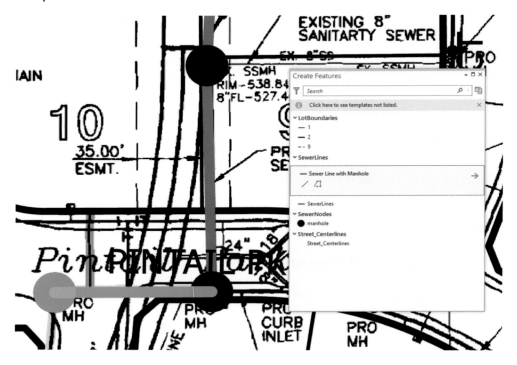

As before, the number of clicks needed to draw features has been greatly reduced, and the possibility of creating errors has been all but eliminated.

YOUR TURN

Zoom to the Sewer Segment 2 bookmark, and add the sewer line as shown. Note that this one has a curve to it, so you will need to use some of the other layout tools, such as the tangent curve, when creating it.

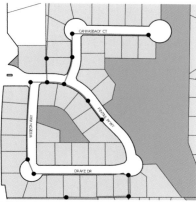

Then zoom to the Sewer Segment 3 bookmark and complete the lines. Make sure you end the lines at the appropriate places to add all the associated manholes. The last segment will be connecting two existing manholes, so you should use the regular SewerLines template instead of the group template to avoid duplicating the last manhole.

If you are feeling ambitious or want more practice with the group template you built, move to the Sewer Segment 4 bookmark, and add the sewer lines and manholes along the south side of Pintail Pkwy and along the west side of Canvasback Ct.

Turn off the scanned image and turn on the Ownership Parcels to see the completed lines more clearly.

17. Save your edits as well as the project.

The small amount of time you spent creating that group template was well spent since now you can drop in sewer lines and associated manholes at lightning speed.

Create complex group templates

In the next scenario, you will create a group template for adding a set of fire hydrant components. When the public works crew adds a fire hydrant, it will always consist of a 6-inch-diameter pipe that taps into an existing water line, a fire hydrant, and a hydrant control valve that is 15 feet from the tap. And although the control valve is always 15 feet from the tap, the hydrant connection pipe can be any length, and it may include bends. Because the length of the pipe can vary, fire hydrants are a better candidate for a group template than a preset or rubber stamp template.

Think for a moment how this set of features might be drawn without using a group template. The line would be drawn, then the control valve snapped to it (possibly with a 15-foot offset from the start of the line), and then the hydrant would be added at the end of the line. Based on this process, can you determine which should be the primary geometry? With two of the three features depending on the line for their location, it's clear that the 6-inch-diameter pipe should be the primary geometry. Then the hydrant can automatically be added at the end of the line. The control valve can also be added automatically using an offset from the starting point of the line. Review the template builders chart to see which builders to use. It should be clear then how to proceed with the group template creation.

1. **Move to the Water System map, and zoom to the Water Lines bookmark.**

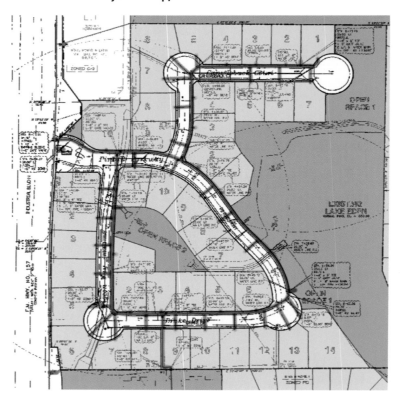

2. **Open the Manage Templates pane.**

3. **Start the creation of a group template for the Hydrant Leads layer. Refer to the storm culvert scenario under "More group templates" if you need help remembering the steps.**

4. **Name the template** Fire Hydrant Set**, provide a description of** Add a fire hydrant lead, a fire control valve, and a fire hydrant**, and add a tag for** Oleander.

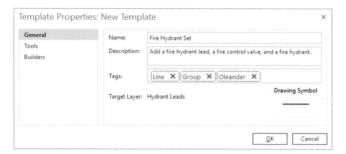

5. Click the Tools tab. Clear all checked tools except the Line tool.

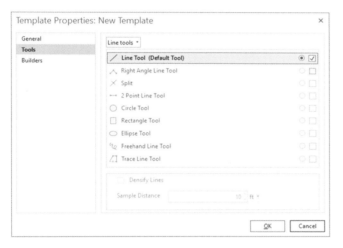

6. Move to the Builders tab, and click Add.

7. Change the feature template to Fire Control Valve, and set "point at beginning of line" as the builder.

8. In the optional "Enter a position along a line" builder, check Distance, and set the distance to 15 (ft). When all the settings are complete, click Add.

9. Click Add again for the final feature. Set the feature template to Fire Hydrant, and select the "point at end of line" builder. Click Add.

10. Click OK to create the template. Close the Manage Templates pane.

Several hydrants are installed with this subdivision, and they are marked in red in the scanned and georeferenced plan. Plus, two additional hydrants were added after these plans were approved at the request of the Oleander fire marshal. There are bookmarks for each of these hydrants, so you can move to the locations and add the new hydrants using the group feature template.

11. Zoom to the Hydrant 1 bookmark.

You can see in the plans the notation showing where to insert the fire hydrant.

12. Open the Create Features pane.

Before you start editing, some attributes must be set for each of the components of the Fire Hydrant Set template. By setting these attributes in the template, they will populate automatically when a feature is drawn. You will need to set the box number to 109 and the year of construction to 2019 for each feature in the set.

13. In the Create Features pane, locate the Hydrant Leads layer and select the Fire Hydrant Set template. Click the Options arrow on the right of the template.

In the Active Template options pane, the three templates that are used to form the group template are displayed. Selecting each one individually will show the attributes for the features and allow you to preset them before any data is created.

14. Make sure Hydrant Leads is selected. Change the year_const value to 2019.

15. Select the Fire Control Valve template, and set the BOXNO value to 109 and the Year of Construction value to 2019.

16. Select the Fire Hydrant template, and also set the BOXNO value to 109 and the Year of Construction value to 2019.

17. Click the back arrow to return to the Create Features pane.

It seemed as if there were a lot more steps to creating this template, but be aware that there are so many more pieces of data associated with this template than the others. Not only will this template draw the set of features with minimal mouse clicks, but it will also populate the attribute table. All these controls lead to better data integrity.

18. Snap to the DistLateral feature at the hydrant location, and draw a line perpendicular and about 35 (ft) long.

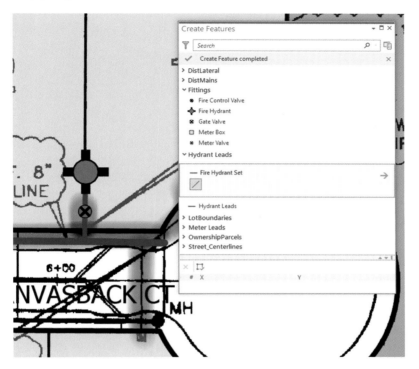

Once again, it took only two clicks to add this set of features.

YOUR TURN

Use the Hydrant 2 through Hydrant 5 bookmarks to add all the additional hydrants that these plans call for. Note that at location 4, the hydrant should be placed on the far right of the median. You may want to use a curve or multipart polyline for this hydrant lead. Then at location 5, you should use a right angle and extend the hydrant lead to the south to show that it doesn't cross the property line.

You will be amazed at how efficient these group templates are and at how much time they will save. As always, the controls over data integrity are extremely valuable.

19. **Save your edits and the project file before continuing.**

Create a preset template

The group templates you've created so far require some sort of user interaction to create a primary geometry from which other template geometries are built. There are a few restrictions on these types of templates, mainly that you can draw only one primary geometry feature and everything must have a snapping relationship with that feature. You saw that drawing a line could produce additional points along the line, and if you look at the template builder rules, you will see that features must always have a snapping relationship.

The preset template is different in that it doesn't have a primary geometry and can produce any number of features, even if they don't have a snapping relationship. In fact, the features wouldn't even have to touch. They are made by identifying and selecting features that are part of a system you want to duplicate, and then converting them into a template. Then as with other templates, you can set any parameters of the included features before placing the item.

When preset templates are created, one important thing to add is an attachment point. This point is where the features will snap to other existing features to ensure proper placement. For instance, if you are adding pipeline features and need to maintain connectivity to the network, the preset template features must snap into the network correctly.

In this scenario, the city is installing new devices for the storm water drainage system that will help remove floating trash from storm water runoff. The devices, named Bag Blasters, have a cyclonic chamber that will separate floating plastic such as plastic shopping bags and plastic water bottles. The items are then dropped into a secondary chamber where they are trapped until the crews come around and clean them out. These items get recycled rather than swept through the creeks as litter.

This new subdivision is getting Bag Blasters, and they are represented as a rectangle for the concrete box, a circle labeled with an *S* for the separation chamber, and two circles labeled with a *C* for the cleanout ports. Imagine for a moment how you could create this set of features manually. You could draw a rectangle for the box, and then add a point for the separator at the centroid of the box. But the two points representing the cleanouts don't have a locational relationship with the other features except that they are inside the box. It would be hard to have these cleanouts in perfect alignment if you always drew them freehand. And it would be impossible to place them using a standard group template. The preset, or rubber stamp, template will be a good option for placing these features as a single system.

The process starts with selecting the features that will be part of the system.

1. **Move to the Storm Drain System map.**

2. Zoom to the Bag Blaster bookmark.

3. Select the four features that make up the Bag Blaster (the polygon and the three points).

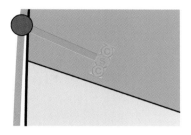

When you create the preset template, you will want the attachment point to be along the edge of the box, just above the separator. These devices attach to the end of the pipe, not over the top of them. The default snapping point will be the centroid of the selected features, and you will need to move that centroid to the correct position. To set the attachment point, you will activate the Move tool and drag the attachment point to the end of the existing pipe. Then use the tools to create the template.

4. On the Edit tab, select the Move tool from the Tools palette. The Modify Features pane opens, and a yellow dot appears, representing the attachment point.

5. Place the cursor over the yellow dot, and press and hold the Ctrl key.

6. Drag the dot to the end of the pipe, and snap it to the Bag Blaster Areas: Vertex. Let up on the mouse button and the Ctrl key, but do not close the Modify Features pane.

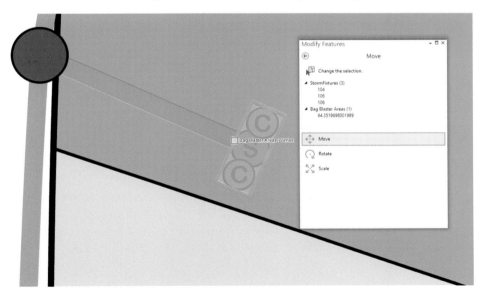

7. Open the Manage Templates pane, and select the Bag Blaster Areas layer. Then click New > Preset Template from Selected Features.

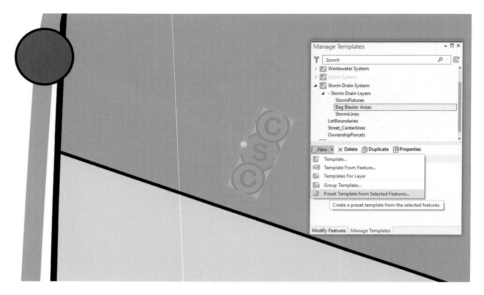

8. Name it Bag Blaster System, and give it an appropriate description and tags.

9. Move to the Features tab, and set the Year_Built attribute to 2019 and the PrivateLine attribute to Yes for all the point features.

10. When you're finished, click OK to create the preset template. Close the Manage Templates pane and the Modify Features pane.

As you can see, preset templates are no more complex to create than other group templates, and in some ways easier because you don't have to define relationships between the features. The only trick is in moving and setting the attachment point for the new template. Now it's time to try out the new template.

11. Zoom to the BB-1 bookmark.

12. Open the Create Features pane, and select the Bag Blaster System template.

13. Place a bag blaster at the end of each of the pipes, and rotate them perpendicular to the pipe. Note that the rotation point will not be the template attachment point, and you will have to move it to rotate correctly.

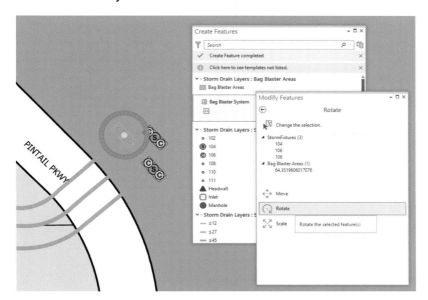

14. Zoom to the BB-2 bookmark, and then place and rotate three more bag blasters.

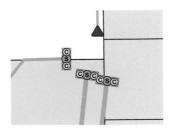

With thousands of other bag blasters being installed in Oleander, this rubber stamp template will get a lot of use in the next few years.

15. Save your edits and the project.

16. If you are not continuing to the next exercise, exit ArcGIS Pro.

Exercise 6-2

The tutorial showed how to build and use various group and preset templates.

In this exercise, you will create two new group templates and use them to add features to the project. The first template will be to add the meter connections for each piece of property in the subdivision. These meter leads will connect to the water main, then run into the yard, and have a meter on the other end. In addition, the meter leads have a meter valve that is 10 feet before the meter. The Landing-Water-Rectified.tif layer in the Contents pane shows these meter leads in pink.

The second group template will be one to draw the lake as a polygon and a lake boundary as a line. You will need to decide which is the primary geometry for these templates, and then construct and use them.

- Start ArcGIS Pro, if necessary, and open Tutorial 6-2.aprx.
- Move to the exercise 6-2 map.
- Use the MeterLeads and Fittings layers in the WaterDistribution_Data feature dataset to create the new template for meter leads.
- Draw all the meter leads using the new template.
- Use the Bodies of Water and Creeks layers to create the new template for the water boundaries. Note that you will need to preset the attribute value for the Creeks layer to symbolize the edge of the water.
- Draw the lake and lake boundary using the new template.
- Save the results when completed.

WHAT TO TURN IN

If you are working in a classroom setting with an instructor, you may be required to submit the maps you created in tutorial 6-2.

- Printed 11 × 17 maps or screenshot images of:
  - Tutorial 6-2 Water, Sewer, and Storm Drain templates
  - Exercise 6-2 Water Distribution and Lake templates

Review

In this tutorial and exercise, you explored creating and using group and preset templates. The ease with which these templates are created pays dividends in both speed and accuracy. Features that may take many mouse clicks to add and configure can be reduced to just one or two clicks. And with the features being configured in advance, there is less opportunity for mistakes, ensuring good data integrity.

The group templates prompt the user to draw the primary geometry, and then the template can draw the rest of the features automatically. There is no limit to the number of features that can be included in the group template, although the features must have a locational reference to each other. Multiple template builder rules can help define them, such as at line vertices, at the centroid of a feature, or on the perimeter of a polygon.

Preset templates can include any number of features, yet the features do not have to have a definable locational relationship. They are made by selecting a set of existing features of any geometry type and then placing them into a preset template. The template is used like a rubber stamp to place duplicates of the features in any position. Care must be taken to set the attachment point of the template to both control placement and ensure data integrity.

With any of the templates, once the features are placed, they go into their origin feature classes and are treated as regular features from then on. If the features must be moved or rotated later, they will no longer have any direct association with each other beyond their connectivity and can be edited the same as regular features.

STUDY QUESTIONS

1. How can you determine the primary geometry of a group template, and what does that mean to the template builder and the user?
2. List advantages and disadvantages of using the preset template.
3. Describe some of the locational relationships that must exist between features in a group template.

Other study topics

Search for these key phrases in ArcGIS Pro Help for further reading:
1. Introduction to feature templates
2. Create a feature template
3. Configure a group template
4. Create a preset template

Chapter 7

Working with topology

One basic concept that distinguishes geographic data from other types of data is topology. In topology, all data with a geographic reference also has a spatial relationship with other geographic data. The data may be adjacent to other data, overlap other data, intersect other data, or be purposefully segregated from other data. The topological relationship can be measured and quantified and can become a permanent characteristic of the data. For instance, the ownership of parcels of land is often represented by polygons, with each polygon containing data about the owner. This dataset is continuous, meaning that all land must be owned by someone. In this model of landownership, there cannot be gaps between the polygons because that would represent land that no one owned. Likewise, each piece of land would have adjacent land with adjacent landowners. The relationship of always having adjacent polygons and never having gaps between polygons is a topographic relationship. In a situation in which these types of topographic relationships are absolute (i.e., always and never), you can begin to use them to your advantage when you edit data.

If two parcels of land are not allowed to have a gap between them, you can create a topology editing tool that won't allow that gap to be drawn. When the topology editing tool is used on one parcel, the tool can automatically apply changes to adjacent parcels to prevent gaps from being formed. Topology has many functions other than making editing easier, but that's the aspect of topology that this chapter explores.

Tutorial 7-1: Working with map topology

One of the most important concepts in the field of GIS is the relationships among features. These relationships may include coincidence, adjacency, separation, and exclusion, to name a few. The geodatabase has components to help construct and maintain the relationships and build a topological fabric. The first relationship to examine is a map topology, which is a temporary topology built on the fly for a specific dataset.

LEARNING OBJECTIVES
- Set up shared edits
- Build a map topology
- Edit features
- Create new features

Introduction

Maintaining the integrity of your GIS data is one of the most important tasks you face. Some datasets have a natural geographic relationship that must be preserved, such as lot lines being at the edge of parcel boundaries. There may be other relationships as well, such as points representing power poles that are attached to lines representing transmission lines.

Topology is the concept that geographic features share a spatial relationship with each other. This relationship may be an adjacency, an overlap, an inclusion, or a coincident edge, among other things. These relationships are critical when doing GIS analysis such as overlays or proximities. When you edit, you must be sure not to interrupt this topological relationship. If one of the features is moved, the other features that share its topology must move with it, even if the original feature is a different feature type in another layer.

One of the simplest types of topology used to maintain this relationship is the map topology. Using map topology, you set up a temporary relationship among several layers in your map that will allow you to do simultaneous edits of these layers. A map topology is created on the fly within a map and is preserved only for that map. If the same layers are added to a different map, the map topology will not be there and must be created again if the same relationship is desired. There are also no user-established rules with a map topology. A basic relationship of coincidence is established between the edges and nodes, which also includes the polygons they may create. This coincidence of feature components makes the map topology easy to configure and use.

Another advantage of the map topology is that it can be applied across geodatabases, and even across file types. Feature classes, whether they are stand-alone or in a feature dataset, from multiple geodatabases as well as shapefiles can all participate in, and be edited by, a map topology.

Once a map topology is created, different topology tools become available for editing that can change features but maintain the topological relationships. The tools can be used for things such as shared edits or even be used to create new features to restore the topological structure.

The other type of topology covered in this chapter is geodatabase topology, which has a rich set of rules for defining relationships among points, lines, and polygons. Geodatabase topology is covered in tutorial 7-2.

Setting topological relationships

Scenario

The City of Oleander recently made changes to the city limit line. Two of these changes happened through a joint annexation with adjacent cities, and one was the result of an error that was recently discovered. Because of these changes in the location of the city limit line, the police district layer is now incorrect, because it follows the city limits.

Two layers represent the police districts. One is the polygon layer used for color shading and overlay procedures, and the other is the linear feature layer used to symbolize the boundaries between districts.

You will use two different editing techniques to make corrections to this data. The first technique involves shared edits. As you edit one feature class, the other feature classes that share a topological relationship will be edited at the same time. The second one involves the tools to create features, a process that will restore missing features in the topological structure.

Data

The map includes the two feature classes making up the police districts, as well as the newly edited city limit line.

Tools used

ArcGIS Pro:
- Map topology
- Topology Edit tool
- Topology: Show Shared Features
- Split tool
- Trace tool
- Topology: Construct Features

Create a map topology

1. Start ArcGIS Pro, and open the project file Tutorial 7-1.aprx.

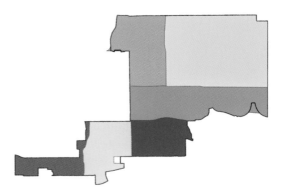

The map contains the new city limit line and the features for the police districts that you'll be correcting. Notice a few areas in which the dark city limit line doesn't match the police district boundaries. These areas are the ones to fix.

As with many of the more complex editing tasks, some setup procedures must be done before editing. It is important to check the status of the Map Topology tools.

2. **Click the Edit tab, and in the Manage Edits group, verify that Map Topology is activated. If not, click the down arrow and select Map Topology from the list.**

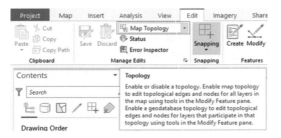

If you modified the police district boundaries without the topology being active, the changes would affect only the linear part of the police districts and would not modify the polygon component of the police districts. You'd have to repeat the process to correct the polygons. Although this work doesn't seem much of a task here, if you had to do it for 20 corrections for a total of 40 operations, you'd want to simplify the task.

Using topology tools to edit features will force ArcGIS Pro to keep the topological relationship intact. If one shared feature is modified—for instance, the police district boundary line—the corresponding feature in the other feature class will also be modified—for example, the police district polygon.

Edit the topology

With Map Topology active, you will be able to make edits on both map layers at the same time and maintain their topological relationship. This example uses two layers, but any number of layers can be included in a map topology.

1. **Move to bookmark PD1.**

Looking at the features, you can get an idea of what the error consists of. The green police district (a polygon layer) has a red boundary (a line layer), and it's supposed to be coincident with the black city limit line. Both the polygon layer and the line layer must be moved so that they match the city limit line. You could edit each one individually, but the topology tools let you edit both layers simultaneously.

Chapter 7 | Working with topology   219

2. Use the Select tool to select both the Police Districts polygon and the PoliceDistricts_Boundaries features that must be corrected.

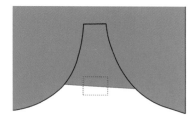

3. On the Edit tab, in the Features group, click Modify, which opens the Modify Features pane.

4. Scroll down the list of tools and click Reshape. Note that the two features you have selected appear in the list of features.

5. At the bottom of the screen, the Tools menu will open. Click Trace.

6. Click on the city limit line below the area where the correction needs to be made, and then trace the line upward and around the city limit line, stopping just past the area where the correction ends. Double-click to end the trace and perform the reshape task.

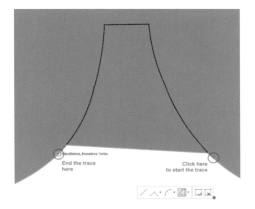

Both the line and polygon layers of the police districts were modified and now align with the city limit line.

7. **Save the edits but keep ArcGIS Pro open.**

   This was a simple edit that required moving a few nodes. The next edit will involve creating new features.

Create features using topology tools

1. **Zoom to the PD2 bookmark.**

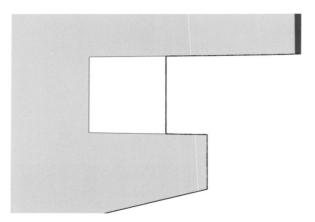

   The bookmark goes to another place where the two layers representing the police districts don't align with the city limit line. In this case, rather than the police district going outside the city limit, the district does not extend fully to the city limit. The solution will be to correct the police district boundary feature, thus causing the topological relationship it has with the police district polygon feature to automatically correct that layer as well.
   Even though this error could be corrected using the Reshape tool, you will use the Vertices tool instead to gain some experience with other topology correction techniques.

2. Make PoliceDistricts_Boundaries the only selectable layer.

3. Select the linear feature in the PoliceDistricts_Boundaries layer along the line where the error exists.

4. On the Edit tab, click Modify to open the Modify Features pane.

5. Select the Vertices tool, and click the Edges tab.

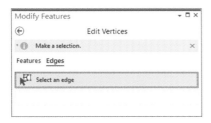

6. Move the cursor over the edge you will modify (the feature will turn purple) and click it (the feature vertices will be shown).

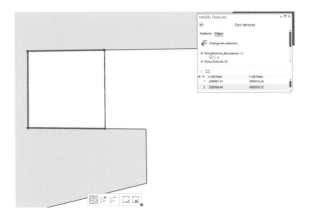

7. Click the upper-left vertex and drag it to the correct position.

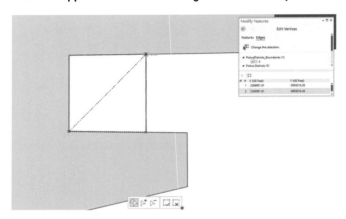

8. Next, click the lower-left vertex and drag it to the correct position.

9. Finish the process by either clicking Finish on the Tools menu at the bottom of the screen or by pressing F2 on the keyboard. Note that the line work and the polygons for the police districts were both corrected.

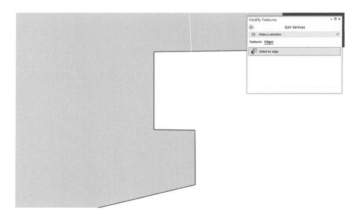

10. Save the edits.

Because the topology was set so that the police district polygons and boundaries must share an edge, you created a situation that broke the topology rule by reshaping the lines to follow the city limit line rather than the police district boundary.

The map topology helped perform these edits by moving shared nodes and automatically creating features that were necessary to keep the topological structure valid.

11. Save your project.

If you are not continuing to the next exercise, exit ArcGIS Pro.

Exercise 7-1

The tutorial showed how to activate a map topology and use it to make edits to the data. In this exercise, you will complete the corrections to the police districts.
1. Start ArcGIS Pro, if necessary, and open Tutorial 7-1.aprx.
2. Zoom to the PD3 bookmark.
3. Activate the map topology for Police Districts and PoliceDistricts_Boundaries.
4. Use the editing tools to correct the discrepancies between the city limit and the police districts. Hint: You can use either the Reshape or Vertices editing tools, or a combination of both.
5. Save the results when completed.

WHAT TO TURN IN

If you are working in a classroom setting with an instructor, you may be required to submit the maps you created in tutorial 7-1.
- Printed 11 x 17 map or screenshot image of:
  - Tutorial 7-1
  - Exercise 7-1

Review

In this exercise, you explored one of several types of topology used in ArcGIS Pro. A GIS topology is a set of rules and behaviors that model how points, lines, and polygons share geometry. Topology helps establish and maintain the relational integrity and behavior among spatial features, which is crucial for a successful GIS. A map topology is one of the simplest types of topology used by ArcGIS Pro and is a temporary topology built on the fly for a specific dataset.

In topology, geographic features share a spatial relationship with each other, whether it is an adjacency, overlap, inclusion, or a coincident edge, among other relationships. These relationships are critical when doing GIS analysis such as overlays or proximities. When you edit, you must be sure not to interrupt this topological relationship. If one of the features is moved, the other features that share its topology must move with it, even if the feature is a different feature type from the others and is in another layer.

The map topology you activated used a temporary relationship between police districts and police district boundaries in your map to allow simultaneous editing of these layers to match the city boundary. This editing preserved the topological relationship for the police districts, police district boundaries, and the city limit line. The topological relationship among these features automatically moved the shared lines and nodes to keep the topological structure valid.

A big advantage of the map topology is that all the layers you want to include in the topology don't have to be stored in the same place, or even be of the same file type. Feature classes and shapefiles, regardless of their source, can participate in a map topology.

STUDY QUESTIONS
1. Why is topology important in a GIS?
2. List the different types of topology available in ArcGIS Pro.
3. What are the benefits of using a map topology, and in what circumstances is it best to use this type of topology?
4. What is the major drawback of a map topology?

Other study topics

Search for these key phrases in ArcGIS Pro Help for further reading:
1. Introduction to editing topology
2. Reshape a topology edge
3. Edit a topological vertex
4. Edit a topology edge

Tutorial 7-2: Working with geodatabase topology

Geodatabase topology is more robust than map topology and is what most people are referring to when they use the term *topology*. Geodatabase topology allows the user to establish which rules to enforce and supplies a rich set of tools for examining and correcting topology errors.

LEARNING OBJECTIVES
- Establish topology rules
- Build a geodatabase topology
- Examine topology errors
- Correct topology errors

Introduction

You learned that map topology was useful for creating a temporary, limited topology within a single map. Now you will look at geodatabase topology. This type of topology is constructed against existing feature classes and stored in the geodatabase for use in multiple projects. Unlike map topology, geodatabase topology persists beyond the map in which it is created.

Geodatabase topology can model the behavior of features represented as points, lines, and polygons. There are also rules for like features, such as polygon-to-polygon behavior. ArcGIS contains a set of predefined rules to manage these relationships, as illustrated in the ArcGIS Geodatabase Topology Rules poster. This guide can be found in the materials provided with this book in ArcGIS Online. The point rules are shown in purple, the line rules in green, and the polygon rules in blue.

The first step in building topology is to determine what relationships exist in your data. Think about a street network as an example. One rule might be that each street segment should be drawn only once, and it should not have a bunch of lines overlapping. A review of the topology rules poster shows two linear rules stating that lines "must not overlap" and lines "must not self-overlap." Read the descriptions of these rules to determine whether either or both might be useful in a street centerline application. Roads also must not go down rivers, so perhaps the rule "must not overlap with" could be applied to streets and rivers.

After a set of rules is determined for the topology, you must gather all the feature classes that will participate in the topology and place them in a single feature dataset. It's a requirement that all feature classes that are to participate in a topology must be in the same feature dataset.

Finally, the topology can be built. Geodatabase topologies are saved as an independent item in the feature dataset and will persist across many projects. They can also be updated later to add or subtract rules or layers.

When the topology is built and validated, a new set of feature classes is created. These special topology feature classes hold indicators of all the features that violate the rules you have established and are used to make corrections. Options to correct the topology errors include manually fixing the error and revalidating the topology, using the topology tools to perform shared edits on the features, or using the topology corrections procedures to automatically fix the errors. The severity of the error and the features involved will dictate which methods can be used.

Among the poster's topology rules for polygons and lines, shown in yellow, is the "must be larger than cluster tolerance" rule. When the topology is built and validated, it will build topology relationship features using your existing features and may not always be able to match the precision of your data. For instance, if two polygons are coincident along a curve, even endpoints that fall within the acceptable tolerance for the point locations cannot produce identical curves along the edge of the polygon. The discrepancies in the topology features will be measured in thousandths of an inch but will show up as a topology error. To prevent having to deal with these miniscule errors, the cluster tolerance rules are mandatory for both lines and polygons. Vertices or edges that fall within the specified cluster tolerance are determined to be coincident for purposes of the topology and will not display as an error.

Setting up geodatabase topology rules

Scenario

You worked with the City of Oleander parcel data in previous tutorials to design the geodatabase and perform some editing. In the future, you may be the one to maintain this data, so you will need to establish a set of topology rules that can help with the task.

Data

Examine the feature classes in the PropertyData feature dataset in the Tutorial 7-2 project geodatabase. Reference the topology rules poster to select rules to use on the following features:
- Blocks: polygons representing sets of parcels for a specific block in a subdivision
- LotBoundaries: lines representing the parcel boundaries
- Parcels: polygons representing pieces of property

Tools used

ArcGIS Pro:
- Create Topology
- Add Feature Class to Topology
- Add Rule to Topology
- Validate Topology
- Error Inspector

Select topology rules

1. Go to ArcGIS Online, and open the topology_rules_poster from the provided course materials for the Focus on GDBs in ArcGIS Pro (Esri Press) book group in the Learn ArcGIS organization. Starting at the upper left of the diagram, read the rules and evaluate how each rule applies to the land records data. Think also of how editing one of the feature classes might affect another.

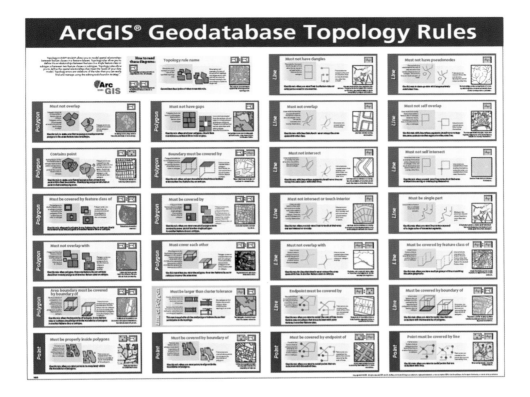

The following suggested set of rules will ensure that property polygons have no gaps or overlaps and that each property boundary is represented by a polygon edge. It also ensures some data integrity in the lot boundary linework. These lines cannot have a dangling end (either an undershoot or overshoot), cannot cover each other, and cannot trace back onto themselves. If you have ideas of other topology rules that you wish to use, add them to the list. Choose carefully because some rules may conflict with each other and create topology errors that can never be fixed. You may see two or more rules that seemingly do the same thing, but there may be subtle differences in their application. Read the full description of each one to fully understand the consequences of using the rule.

Polygons:
- Must be larger than cluster tolerance—this mandatory rule will prevent tiny discrepancies in the topology features from showing as feature errors.
- Must not overlap—for the Parcels feature class, you can't have overlapping polygons, or it would mean that two people own the same piece of property. Other features that are mutually exclusive are the blocks.
- Must not have gaps—the parcels must not have gaps between them, or there would be property that no one owns. Since the blocks occur only on platted property, they can have gaps.

- Boundary must be covered by—the Parcels features must be covered by a Lot Boundaries feature. Use this rule when you intend to edit the polygons and have the line work automatically adjust. The interesting thing about this rule is that it can be set to respect subtypes in the data. For the Blocks feature class, you will set a rule for only the subtype of Platted Property.

Lines:
- Must be larger than cluster tolerance—this mandatory rule will prevent tiny discrepancies in the topology features from showing as feature errors.
- Must not have dangles—the line features in the Lot Boundaries feature class must enclose polygons, so they cannot have lines with ends that do not snap to another line.
- Must not overlap—the Lot Boundaries feature class should not have multiple features representing the same lot line. Having features that overlap creates a problem with both editing and symbolizing.
- Must not self-overlap—it is never desirable to have a line feature double back on itself. Set this rule for the Lot Boundaries feature class.
- Must be covered by boundary of—the lot boundaries must be covered by the edge of a parcel feature. The intent here is to edit the line work and let the polygons be automatically created. Also, there is a one-to-one relationship between the lot boundaries and the parcel edges.

Note that some of these rules apply only to a single feature class, whereas others establish a relationship between two feature classes, sometimes of different feature types. In your evaluation of the rules, you may have decided to add a few not listed here. This addition will make the rules more stringent, but be careful not to apply two rules that either do the same thing or will conflict with each other.

Creating the topology and setting it up for use involves four steps:
1. Create the topology structure. Remember that this structure can exist only within a feature dataset, and all feature classes that will participate in this topology must also be in the same feature dataset.
2. Add feature classes to the topology. One by one, you will specify which feature classes will participate.
3. Set a feature rank in the topology. For instance, because the lot boundaries for Oleander are entered from survey data, they are the most accurate and are used to later create the parcels. The lot boundaries should have a higher ranking than the parcels. Conversely, the block data is the least accurate and is merely a reference dataset, so it can have the lowest ranking.
4. After adding feature classes to the topology, add the rules.

Build the topology

1. Start ArcGIS Pro, and open the project Tutorial 7-2.

2. Open the Geoprocessing pane, and search for the Create Topology tool. Click the tool to open it.

3. Set the input feature dataset to PropertyData within the Tutorial 7-2.gdb geodatabase. Name the output topology Oleander_Topology. Set the cluster tolerance to .02, and click Run.

4. Search for and open the Add Feature Class To Topology tool.

5. Set the input topology to the one you created in step 3 (Note: You can use the Browse button to select it, or drag it from the Catalog pane). Use the drop-down arrow to set the input feature class to Lot Boundaries. Because this feature class is the source data for the parcel polygons, leave its XY rank at 1. Click Run.

In the topology, each layer must have an XY rank and a Z rank (although this data won't need a Z rank), in which 1 is the highest. The layers with the highest rank will not be altered in an automated topology correction process. Layers with lower ranks will be adjusted to be coincident with the higher ranked layers. In this example, the lot lines are the highest ranked since they are typically entered from accurate survey data.

YOUR TURN

Using the Add Feature Class To Topology tool, add Parcels to the topology with an XY Rank of 2. When you're finished, add Blocks to the topology with an XY rank of 3.

These ranks will mean that the lot boundaries created from survey data will never move. But parcels, polygons created from the lot boundaries, may be adjusted to match the lot boundaries when needed. And the Blocks polygons can be changed however necessary since they are an abstract representation of a characteristic of the plat process.

Verify the results by going to the Catalog pane and opening the Properties pane for Oleander_Topology, and then click the Feature Class tab. Close the Properties pane when you have examined the results. All the feature classes in this project are involved in this topology, but if things were to change, you could remove feature classes with the Remove Feature Class From Topology tool.

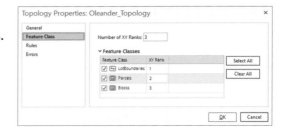

---

6. Search for and open the Add Rule To Topology tool.

Reference the list created earlier for the rules to assign to this topology. Note that there is a mandatory rule of "must be larger than cluster tolerance" for each layer. You already set the cluster tolerance when the topology was created, so this rule is added and maintained automatically, even though you won't see it later in the applied rules list.

7. In the Add Rule To Topology dialog box, set the input topology to Oleander_Topology. Then set the rule type to Must Not Overlap (Area) and the input feature class to Parcels. Click Run.

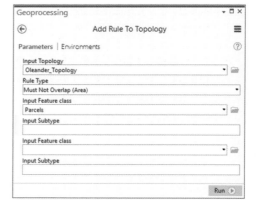

Note that there are two places to enter a feature class name. In this instance, only one feature class was entered, which means the polygons of this feature class cannot overlap themselves. But if a second feature class was entered, it would apply this topology rule across both feature classes and not allow them to overlap each other.

8. **When the process completes, reconfigure the tool to add the Must Not Have Gaps (Area) rule to the Parcels feature class. Click Run.**

    Again, note that this is a self-protection rule. It will keep polygons within the feature class from having gaps.

9. **Change the tool to add the rule type Must Be Covered By Feature Class Of (Area-Area). Set the input feature class to Parcels with the input subtype of** Platted Property. **Then set the second input feature class to Blocks. Click Run to add the rule.**

    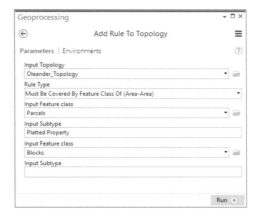

    This rule acts on both features classes to make sure that blocks can exist only where there is a parcel. Further, by adding the subtype value, it says that blocks can exist only where there is a platted parcel. Because unplatted property doesn't have a block designation, leaving out the subtype reference would mean that all unplatted property violates the topology rule.

10. **In the Catalog pane, open the properties for Oleander_Topology. Click the Rules tab to verify that all three rules have been added. Click OK to close the dialog box.**

This dialog box also has tools that let you add new rules, remove rules, save a set of rules, or load a previously saved set of rules. The Topology Properties pane also displays general information about the topology, including its validation status, the feature classes involved in the topology, and the errors found after validation.

YOUR TURN

Add the following rules to Oleander_Topology. Use either the Add Rule To Topology tool or open the Topology Properties pane, highlight the Rules tab, and use the Add button to add rules to the list as follows:

Feature Class 1	Rule	Feature Class 2
Blocks	Must Not Overlap (Area)	
LotBoundaries	Must Not Have Dangles (Line)	
LotBoundaries	Must Not Overlap (Line)	
LotBoundaries	Must Not Self-Overlap (Line)	
LotBoundaries	Must Be Covered by Boundary Of (Line-Area)	Parcels

Match these rules to the topology poster if you need a better understanding of what each is doing.

When you're finished, view the topology properties and verify that all eight rules were added successfully. Save the rules to the project folder for future reference.

The final step in creating the topology is to validate it, which will check the existing data against the new topology rules you created. This first time through, there will be a lot of topology errors that must be fixed. In the future, you would have to fix topology errors only after editing.

The Validate Topology tool will, by default, run the validation checks on the current display extent. Any time you want to validate the entire dataset, you must zoom the map to its full extent.

11. Zoom the map to full extent.

12. Find and run the Validate Topology tool using Oleander_Topology for Input Topology.

Examine the linear errors

When Oleander_Topology was created, a special group layer was added to the Contents pane. It contains topology feature classes to show and identify topology errors. The Dirty Areas layer shows places where corrections to the topology have been made and need to be revalidated. The other three topology layers show the features that are contributing to the error. The goal is to examine every error and either fix it or mark it as an exception to the rule. This task seems long and tedious, but once your dataset is topologically perfect, the only errors you will have to deal with are those introduced in new edits—and with care, those can be kept to a minimum.

1. In the Contents pane, turn off the Parcels, Blocks, and Polygon Errors layers. This setup will allow you to concentrate on fixing linear errors first.

2. Click the Edit tab, and in the Manage Edits group, use the drop-down box to identify which topology you want to work with. Select Oleander_Topology (Geodatabase).

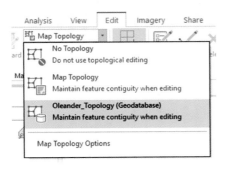

3. Also in the Manage Edits group, click Error Inspector to open the Error Inspector table, from which you will manage the corrections. Make sure the filter button for Map Extent is not checked so that all the errors will display. Then click to examine the first few errors. You can click in the Preview area and zoom out a little to get a better view of the errors.

The Error Inspector table displays all the topology errors that were found in the data. You can show all the errors or just a subset of errors. Each error can be selected and will be displayed in a viewer on the right of the pane. Clicking Fix along the top of the viewer will present several automatic fix tools for the error. Also along the top of the viewer are tools that will let you restrict which type of errors are shown and validate the topology once corrections have been made.

You will start with the Must Be Larger Than Cluster Tolerance error since these errors can all be marked as exceptions. These errors are segments that are smaller than the cluster tolerance but are not errors in how the data was drawn, so they are all right to keep.

4. **At the top of the Error Inspector table, click Filter . In the resulting list, clear the check box next to Show Exceptions and select Must be Larger Than Cluster Tolerance.**

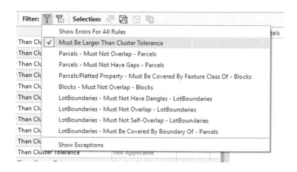

5. **If you want to get a feel for what these errors are, click a few of them to see the features that are creating the error. When you are finished viewing them, select all the errors. On the right side of the viewer, click Fix, and then click Mark as Exception.**

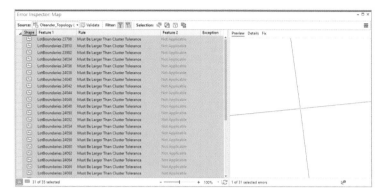

All the errors are corrected as a batch and disappear from the Error Inspector table.

For other error types, the Error Inspector tools may give you options to correct multiple errors at once. You may need to extend a line to meet the boundary, trim a line that has gone past its boundary, or snap the end of a line to the boundary. There may also be circumstances in which the lines are drawn correctly yet do not need to touch the boundary.

6. **Change the filter to Lot Boundaries - Must Not Have Dangles. Select a few of the errors, and click the Preview tab to see the issue. You will see that the lines either fall short of their intended snapping target or run past it.**

7. **Select all the errors, and click the Fix tab. Select the option to Extend with a maximum distance of** 10 (ft), **and press Enter.**

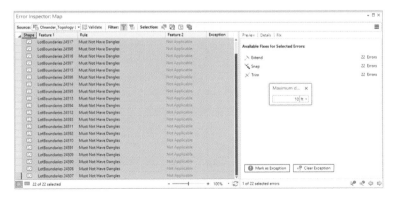

Note: If you get an error that the lines cannot be fixed, make sure they are available for editing on the List By Editing tab on the Contents pane.

8. Repeat the process for the remaining Must Not Have Dangles errors, except this time use the Trim fix at 10 (ft).

9. The fixes that you applied will have modified the features, so save the edits before continuing.

10. In the Contents pane, turn on the Dirty Areas layer. Blue polygons are shown around each feature that has been corrected using the error correction tools but has not yet been validated.

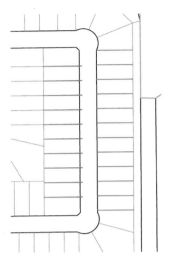

These blue areas, known as Dirty Areas, show areas where features have undergone some sort of change since the topology was last validated. When you run the validate process, it matches those areas against the topology rules to make sure that the changes both solve the former error and do not unintentionally create new errors.

11. On top of the Error Inspector table, click Validate.

For these errors, the Error Inspector tool presented three suggested methods for correcting the errors. The solution was to try to either trim, extend, or snap the lines until they intersected the closest linear feature—or mark them as an exception. This is the same task performed by the Extend and Trim tools, except that the topology rule allows the corrections to be done in a batch process that corrects all the lines at once. It's advisable to try to correct the errors with the Trim and Extend tools first. These tools will modify only the feature that is reporting the error. The Snap correction, however, may move other adjacent features that are not currently causing the error.

Next, you will try an interactive error-fixing tool. At the bottom of the Error Inspector viewer is the Error Correction tool.

Selecting a topology error will present a pop-up menu with possible fixes. Note that this is a one-at-a-time method as opposed to the batch method used before.

12. Change the Filter: Rules to Lot Boundaries - Must Not Overlap.

13. Click the first error, click the Error Correction tool, and select the first error in the viewer. It will be highlighted in yellow.

14. In the pop-up box, click Remove Overlap.

You are then prompted to decide which part of the line to keep. Clicking each choice will highlight the segment in the map.

15. Select the one marked (Largest), and click Remove Overlap.

YOUR TURN

Continue through the list of Must Not Overlap errors, and use the Error Correction tool to remove the overlap. You can use the Zoom To button to find the error in the map. If the Dirty Areas layer is still visible, you will see the blue polygons pop up as each correction is made. When you're finished, validate the topology and save the edits.

Then change the Filter: Rules to Must Not Self-Overlap. Read the choices of fixes and decide which one should be used to correct the errors. When all the errors are corrected, validate the topology and save the edits.

The last of the linear topology errors are for the rule Must Be Covered By Boundary Of. All these errors are created when two of the ROW polygons touch. These errors exist only because the ROW polygon would be unwieldly if it was left as a single polygon for the entire city, so it was forcibly split into smaller polygons for convenience. Use the Error Inspector to select all these errors and mark them as exceptions. Make sure to validate the topology and save the edits when you are finished.

As a final check, change the filter to Show Errors for All Rules, and make sure that there are no errors in the LotBoundaries layer. If there are, select and correct them.

This process will finalize all the linear topology errors. They were corrected first because the accuracy of the linear features is higher, and they are used to create the polygon features.

Examine the polygon errors

The remaining topology rule errors are polygon errors. Different sets of correction methods are used for these errors. Some may require manually moving vertices or adding features, and some may allow for an automatic or batch fix.

1. Turn on the Parcels layer and the Polygon Errors layer.

2. In the Error Inspector table, change the filter to Parcels - Must Not Have Gaps. Click on the first error, and it will highlight in yellow in the viewer.

This error is reported because the polygon representing the city boundary has a gap between it and the rest of the universe. There's no correction for this error but to mark it as an exception.

3. Right-click and set it as an exception.

4. Click the next error.

This error has happened because survey data was entered incorrectly. The parcel below the error should include the gap area. The fix will be to create a feature to correct the error, and then merge the new feature with the feature below it.

5. Right-click the error, and click Create Feature.

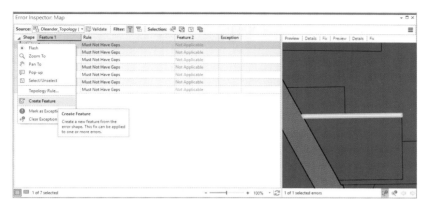

You've worked with editing features before, so the instructions to fix this error are more concise.

6. To merge the two features:
   - Select both features.
   - On the Edit tab, look on the Tools menu and open the Merge tool.
   - Under Features to Merge, select the one labeled ECTOR—it will add "(preserve)" to the description.
   - Click Merge.
   - Finish by deleting the Lot Boundaries feature that was at the edge of the gap.

This process is standard for fixing gaps.

YOUR TURN

Continue going through the Must Not Have Gaps errors. Most are the same type of error, in which you can create a new feature and merge it with one of the existing features. The correct solution will be apparent as you fix the errors.

The last two errors are different from the rest. On the error shown, you will need to create the new feature, and then split it to match the parcels to the west. Finally, merge each piece to the existing parcels.

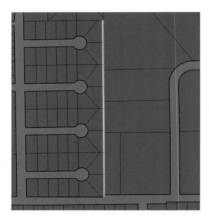

The final error is a section of right-of-way that was overlooked. It has no address or legal description. The best solution is to create the feature, and then set the use code to PROW (Private Right-of-Way).

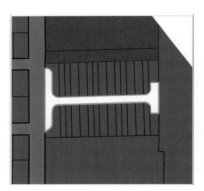

Next, you will look at the features that violate the rule Must Not Overlap. There are a few reasons that polygons might overlap. Some will have survey errors, the same types of mistakes that caused the gaps you fixed earlier. Some may represent a doughnut polygon for which the doughnut hole wasn't removed. But the most common cause of errors you will see in this dataset involves condominium parcels. In this instance, the same piece of property is owned in shares by several people, and because each parcel has a unique tax ID, the polygons will appear multiple times. These errors can be marked as exceptions.

7. Set the Filter to Parcels – Must Not Overlap. Click through and examine the features. Note that the first 160 errors are the type caused by condominiums.

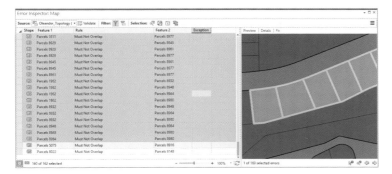

8. Select from the first error through Parcels 8964, and mark them all as exceptions.

9. Select the next error.

This error is a location in which there is supposed to be a median at the entry to this subdivision. However, the gray right-of-way polygon didn't get cut, and it overlaps under the median polygon. The fix for this doughnut hole scenario is to cut a hole in the larger polygon and eliminate the overlap.

10. With the feature selected, click the Fix tab. Then click the Remove Overlap tool to cut out the overlapping piece.

The error on the next feature is a more common overlap. The solution is to use the Merge tool and combine the overlap with one of the polygons and delete it from the other.

11. Zoom the map to the feature, and click the Fix tab if necessary. The entire overlap portion should belong to the parcel to the north.

12. In the Error Inspector Fix box, click the Merge tool. Click the two parcels to see which one is to the north. With it highlighted, click Merge.

YOUR TURN

The last error for this rule is a more standard merge error. Zoom to the error, select the Merge tool, and fix the error.

When all the errors for this rule are corrected, zoom to the full extent of the data and run a validation check and save the edits.

The last topology rule to work with is the Parcels: Platted Property Must Be Covered By Feature Class Of Blocks. The Blocks layer is simply a visual display of block boundaries and is typically made by dissolving parcel polygons by a block number attribute. It's common for blocks to have a lot of sliver errors that can be marked as exceptions, but the topology rule can also identify subdivisions that do not have a block polygon. That's what you will look for in this dataset.

13. Set the filter to Parcels: Platted Property - Must Be Covered By Feature Class Of Blocks. Select and examine each error. If it is a sliver or isolated parcel polygon, mark it as an exception.

14. The parcels with a true error are shown, and many of them are from errors you resolved earlier. Find and highlight these errors. Then on the Fix tab, click Create Feature.

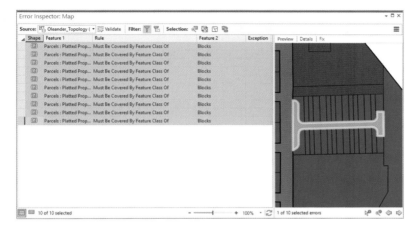

15. Zoom to the full extent of the data, validate, and save the edits.

YOUR TURN

Change the Error Inspector filter to Show Errors For All Rules. The error corrections made to solve other topology errors may have themselves created other errors. Scroll through the list and use the techniques shown in this tutorial to resolve them. When you're finished, validate the entire dataset and save the edits.

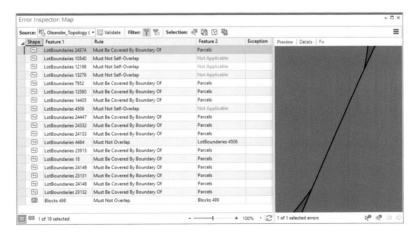

Getting a topologically perfect dataset can be a long and tedious process, but the rewards are that future edits will be easier and the precision of your data will be sharp.

16. **Save your project.**

If you are not continuing to the next exercise, exit ArcGIS Pro.

Exercise 7-2

This tutorial showed how to create and validate a geodatabase topology. It also demonstrated many of the editing techniques to find and fix topology errors.

In this exercise, you will add a new subdivision and practice using the topology tools to fix some of the errors it may create.

A new subdivision named Horseshoe Bend is being added to the City of Oleander. The parcels and lot boundaries have been drawn into new feature classes, which can be copied and pasted into the main datasets. This data is based on a survey that used 21st-century technology and must be placed into an area of the map that was surveyed using 1970 technology. It will undoubtedly create gap and overlap errors.

Follow these steps to integrate the new data:
- Add a new group layer to the Contents pane, and add the HorseshoeBend_Parcels and HorseshoeBend_Boundaries layers.
- Select all the features from the HorseshoeBend_Parcels layer, copy them, and paste them to the main Parcels layer.
- Copy all the features from the HorseshoeBend_Boundaries layer, copy them, and paste them to the main LotBoundaries layer.
- With these new features added, validate the topology.
- Open the Error Inspector table, and review the linear rule errors. Fix them as necessary.
- Review the polygon rule errors and fix them as necessary.
- Validate and save the edits as you progress through the errors until all have been repaired.
- Save the project when completed.

WHAT TO TURN IN

If you are working in a classroom setting with an instructor, you may be required to submit the maps you created in tutorial 7-2.
- Printed 11 × 17 map or screenshot image of corrected features:
  - Tutorial 7-2
  - Exercise 7-2

Review

Of the two types of topology supported in ArcGIS Pro, the most sophisticated is geodatabase topology. This type of topology is constructed for existing feature classes and is stored in the geodatabase for future use. Unlike a map topology, a geodatabase topology persists beyond the original map and can be used in other maps and other projects.

A geodatabase topology can model behavior among points, lines, and polygons. There are also rules for like features, such as polygon-to-polygon behavior. ArcGIS contains a set of predefined rules to manage these relationships. You used some of the polygon and line feature topology rules to complete the exercise. A variety of rules are available to customize how you want ArcGIS Pro to manage the relationships among features.

You first determined the relationships among the blocks, parcels, and lot boundaries. From these relationships, you determined which of the topology rules would accomplish what you needed. You created a geodatabase topology, added feature classes to it, and set up topology rules. All the feature classes needed to exist in the same feature dataset, and feature classes can participate in only one topology.

When building topology, you ranked each of the layers represented in the topology so that the topology editor tools knew which layers should be modified during the process. Typically, you rank the most accurate datasets higher. In this case, you selected the survey-based LotBoundaries layer for the highest ranking.

Once your topology was built, a new set of feature classes was created. These feature classes contained indicators of all the features that violate the rules you established, and the indicators were used to make corrections. By bringing the topology layer into ArcGIS Pro, and reviewing it in the Error Inspector table, you were able to identify features on the map that did not meet the topology rules you specified. You corrected these errors using tools provided by the Error Inspector.

The Validate Topology command was used to ensure data integrity by validating the features in a topology against the topology rules. When resolving errors, you have the option of marking an individual error or a collection of errors as exceptions. There will be instances when the occurrence of a defined error may be acceptable. In the exercise, the roadways created gaps among the Parcels polygons, which violated the Must Not Have Gaps rule. In this case, you marked it as an exception. Once an error is marked as an exception, it remains that way until it is reset as an error. Running Validate Topology will not generate an error for an instance that has been marked as an exception.

Building and maintaining geodatabase topology is difficult, tedious, laborious, time consuming, and generally no fun. Many people choose not to make their datasets topologically perfect. However, the time and effort to build topological precision into your datasets will pay great rewards in the future. Sharing imperfect data online or in a 3D dataset will compound your errors to the point of making them almost useless, not to mention the basic editing issues you will encounter daily.

STUDY QUESTIONS
1. While creating the topology, how can you determine what rules to use?
2. List five geodatabase topology rules from the topology rules poster that were not used in the exercise and to what feature type(s) they pertain.
3. Contrast and compare geodatabase topology with map topology.

Other study topics
Search for these key phrases in ArcGIS Pro Help for further reading:
1. Helpful hints when using topologies
2. Design a geodatabase topology
3. Common topology tasks
4. Validating a topology
5. Geodatabase topology rules and fixes

Index

2-Point Line, 192–93
3D Analyst, 122
3D scenes, 122–31; converting map to scene, 126–29; examination of data, 124–26; exercise, 131; negative extrusion values, 130; review, 131; scenario, data, tools, 123–24; study questions/topics, 131

Absolute X,Y,Z, 145–46
Add Clause, 96
Add Feature Class To Topology, 229–30
address information, 6
address locator, 6–7
Add Rule To Topology, 230–32
alias, 5, 6
Analyze tool, 113, 118
Append tool, 84, 88, 92–93, 95–99, 101, 104
ArcCatalog, 79
ArcGIS®, 1, vii
ArcGIS Enterprise, 107
ArcGIS Geodatabase Topology Rules poster, 225, 226–27
ArcGIS Online: putting data on, 107–22; storage components in, 107–8
ArcGIS Portal, 107
Arc Segment, 149
attachment point, 208, 209
Attributes, 177–78
attributes, automatically populating, 178
attributes template, 59–60, 204–6

Bookmarks, 140

Calculate Field, 89–90, 126
Catalog pane, 38, 41, 83
check box, Ctrl+Click technique for, 156
clause, adding, 96
cluster tolerance rule, 225–26, 230
coded values, 11–12, 13, 16, 24–25, 43

Comments, 99
Configure tool, 144
Construction Tools, Feature. See Feature Construction tools
Contents, 135–40
context menu creation tools, 153–68
copy/paste/move technique, 175–76
Create Feature Class, 68–69
Create Features, 59–61, 73–74, 142–43, 144, 199–200, 204–7, 210–11
Create Topology, 229
Ctrl+Click technique, 156

dashed line, 146
data: See also geodatabase, loading data into; subtypes, loading data into; logical model; manual loading, 79–80; point, loading, 100–102; presenting in 3D scenes (see 3D scenes); steps for designing successful, 76; tabular, 16
data, exercise, 120–21; multiple layers, 114–16; overview of layers and maps, 107–8; preparation of data, 110–14; scenario, data, tools, 109; sharing online, 107–22; study questions/topics, 122; web map, 117–20
data integrity rules/techniques, 9–14, 62; domains, 10–13, 24–25, 42–46; nulls, 9–10, 24; in property parcel geodatabase, 9–14; in sewer line geodatabase, 24–29; subtypes for, 13–14, 25–27, 28–29, 53–54, 62–63, 70–72; on the web, 54; for web map, 118–20
data segregation, with feature datasets, 15–16. See also subtypes
default map extents, 140–42
definition query, 108
Delete Vertex, 146
Direction/Length, 146–47
direction measurements, units for, 171–72
Dirty Areas layer, 233, 236
Discover codes, 71
Distance, 148

domains, 10–13, 24–25, 42–46, 62; assigning to fields, 49–52; coded values as, 11–12, 13, 16, 24–25, 43; contingent, 50–51; creating, from table, 46–48; creating through field design pane, 67; naming, 42; problems with altering, 42–43; range as, 10–11, 24, 43
drawing tools. See Feature Construction tools

edge, 154
edit grid, 176–77
editing, turning off, 140
Enable Undo, 86
endpoint, 153
enterprise geodatabase, 35, 61
Error Correction, 236–37
Error Inspector, 233, 243
Error Inspector table, 233–37, 238–39
Expression, 127
Extend tool, 236
extents, default map, 140–42
Extrusion, 126–27
extrusion, negative values, 130

Feature Builders reference table, 184
feature class, 38–40, 62–63; creating new, 5–6, 69–70; in geodatabase topology, 225; multiple, 93–105
Feature Class To Geodatabase, 92
Feature Construction tools: commonly used (toolbar), 144; for lines, 143–49, 159–66
Feature Construction tools, for polygons, 169–83; buildings, 172–76; edit grid, 176–77; exercise, 181–82; review, 182–83; scenario, data, tools, 170; setting up editing, 171–72; study questions/topics, 183; symbolizing unique features, 177–81
feature dataset, 5, 15, 36–37, 65–70
feature layers, 107–8. See also layers
features: default value for new, 178; display order of, 180–81; with multiple parts, 158

features, shortcuts for creating, 154–68; cleaning up line work, 164–66; edit and trace, 156–60; exercise, 166; multiple parts, 158; parking lot, 160–66; preparing for editing, 160; review, 167; scenario, data, tools, 154–55; setting selections, 155–56; snapping, 157; study questions/topics, 168
feature template, 183–211. See also group template; preset template; table template
field definitions, 68–69
field map, 82–83, 85–86, 95, 99
fields: adding new, 6–9, 38–40; alias for, 6; assigning domains to, 49–52; importing definitions of, 68–69; naming, 6–8; removal of, 70
Find tool, 7

geodatabase, complex, 64–77; adding to data structure, 64–65; creating data structure, 65–68; exercise, 74–75; feature classes, 69–70; field definitions, 68–69; review, 75–76; scenario, data, tools, 64–65; study questions/topics, 76–77; subtypes, 70–72; testing the rules, 72–74
geodatabase, creation of, 33–77; assigning domains to fields, 49–52; building, 33–63; creating domain from table, 46–48; data structure, 34–36; domains, 42–46; exercise 2-1, 61; feature classes, 38–40; feature dataset, 36–37; relationship class, 55–57; review, 61–63; scenario, data, tools, 34; study questions/topics, 63; subtypes, 53–54; testing subtypes, 58–61
geodatabase, goal of designing, 2
geodatabase, loading data into, 79–93; checking results, 87–88; exercise, 91–92; field map, 82–83; linear data loading, 88–91; load procedure, 80–82; review, 92–93; scenario, data, tools, 80–82; starting load process, 83–87; study questions/topics, 93

geodatabase topology, 216, 224–44; adding rules to, 230–32; building, 229–33; exercise, 244; feature classes in, 229–30; linear errors, 233–38; polygon errors, 238–44; review, 245; rules for, 225–26, 227–28; rules poster, 225, 226–27; scenario, data, tools, 226; selecting rules for, 226–28; steps for creating and setting up, 228; study questions/topics, 246; validation of, 232–33
Geoprocessing pane, 229
group layers, 83; adding, 136–40; adjusting scale for, 138; described, 136–37; display settings for, 139; nested, 140
group template, 184, 185–94; complex, 201–8; for drawing line segments, 191–94; exercise, 212; polygon features, 186–91; properties and tools, 189–91; review, 212–13; scenario, data, tools, 185–86; simple, 196–201; study questions/topics, 213

Import, 66
importing: with Append tool, 84, 88, 92–93; field definitions, 68–69; shapefile data, 92–93
Input Datasets, 84, 88, 96, 99
Interpolate Shape, 127
Intersection Snapping, 157, 163

layers. See also group layers: feature, 107–8; map, 107–8; renaming, 136, 137–38; scale ranges for, 138; selectable/not selectable, 155–56; selecting by attribute, 90, 104; turning off editing for, 140; web, 108, 111–13, 115
layers, sharing online, 107–8; feature, 108; multiple, 114–16; web, 108, 111–13
line. See also linear features: dashed, 146; edge of, 154; endpoint of, 153; geodatabase topology rules for, 228; vertex of, 153
linear data, loading, 88–91
linear features, 133–53. See also line; Absolute X,Y,Z tool for, 145–46; bookmarks, 140;

changing symbols for, 149–50; construction tools for, 143–49, 159–66; default map extents, 140–42; Direction/Length tool for, 146–47; errors in, in geodatabase topology, 233–38; exercise, 150–51; group templates for creating, 191–94; parallel, 147–48; review, 151–52; scenario, data, tools, 134–35; in sewer line, 23–27; study questions/topics, 152–53; table of contents, 135–40; Tangent Curve Segment tool for, 148–49
Line: Parallel, 147–48
Line: Perpendicular, 161–62, 164
List By Drawing Order, 149
List By Editing, 140
List By Selection, 155–56
List By Snapping, 157
logical model (point and linear feature classes), 21–29; beginning design process, 23–24; data integrity, 24–29; data structure, 22–23; domains, 24–25; exercise, 29–30; nulls, 24; review, 31; scenario, data, tools, 22–23; study questions/topics, 31–32; subtypes, 25–27, 28–29
logical model (polygon and linear feature classes), 1–19; accessing data, 4; building the model, 4–9; considerations before designing, 2; data integrity rules, 9–14; designing the data, 3–4; exercise, 19–20; relationship class, 16–19; review, 20–21; scenario, data, tools, 3; study questions/topics, 21; using polygon and linear feature classes, 1–19

Manage Subtypes, 71
Manage Templates, 189–91, 191–93, 195, 197–99, 202–4
map: converting scene to, 126–29; field, 82–83, 85–86, 95, 99; web, 117–20
map extents, 140–42
map frame, renaming, 136
map layers, 107–8

map topology, 216–24; creating, 217–18; creating features in, 220–23; editing, 218–20; exercise, 223; review, 223–24; scenario, data, tools, 217; study questions/topics, 224
Mark as Exception, 235
Material query, 998
Merge, 241–42
Metadata, 110–12, 114–15, 138
Microsoft Excel®, 2
Midpoint Snapping, 157
Move, 175, 209

New Group Layer, 83, 136–40
null values, 9, 10, 24

Offset, 156–57
online data, sharing. See data, sharing online
Origin and Rotation, 176
Output Fields, 85

Parallel line tool, 147–48
Paste, 175
Paving, 160
perpendicular line tool, 161–62, 164
point data, loading, 100–102
point features, 27–29
polygon: edge of, 154; endpoint of, 153; errors in, in geodatabase topology, 238–44; Feature Construction tools for, 169–70; geodatabase topology rules for, 227–28; loading, 83–87; vertex of, 153
preset template, 185, 194–96, 208–11; exercise, 212; review, 212–13; study questions/topics, 213
primary geometry, 184

quadrant bearing, 171–72
quality control. See data integrity rules/techniques; subtypes, testing of

range of values, 10–11, 24, 43
relationship class, 16–19, 55–57
Remove Overlap, 237, 241
Reshape, 219–20
Reverse Direction, 162
Right Angle Line, 188
rubber stamp template, 185, 194–96, 208–11

Save Edits, 150
scale ranges, 138
scenes, 122, 126. See also 3D scenes
Schema Type, 96, 99
Select Layer By Attribute, 90, 93–105, 104
Shapefile, 79
Share As Web Layer, 111–13, 115
Share As Web Map, 117–18
Sketch tools, 148, 159–66. See also Feature Construction tools
snapping, disabling, 157
snapping environment and features, 153–54, 157, 171
snapping tolerance, 154
spatial reference, 36, 65–66, 109
Split Feature, 165
spreadsheet for data, 2
Square and Finish, 174
subtypes, 13–14, 25–27, 28–29, 53–54, 62–63, 70–72; creating more complex, 71–72; testing of, 58–61, 72–74, 75–76, 87–88, 100, 102–3
subtypes, loading data into, 93–105; exercise, 103–4; field map, 95; loading data, 95–99; loading data into feature classes, 94–95; loading point data, 100–102; review, 104–5; scenario, data, tools, 94–95; study questions/topics, 105; testing results, 100, 102–3

Survey Control Points, 172
Symbol Layer Drawing, 181
symbols, 149–50, 177–81

table of contents, 135–40
table template, 184–85
tabular data, 16; creating domain from, 46–48; worksheet for, 6
Tangent Curve Segment, 148–49
Tangent Snapping, 157
Target Dataset, 84, 88, 96, 99
template. See also group template; preset template: attributes, 59–60; table, 184–85
topology, 15, 215–16. See also geodatabase topology; Map topology
topology tools, 218, 220–23
Trace, 156–58, 162, 178, 219
transparency level, 139
Trim, 165, 236

Undo, 86
Unique Values, 54, 179
units, for direction measurements, 171–72

Validate Topology, 233
values: coded, 11–12, 13, 16, 24–25, 43; null, 9, 10, 24; range of, 10–11, 24, 43; unique, 54, 179
vertex, 153
Vertices, 163, 220–22

web layer, 108, 111–13, 115
web map, 108, 117–20
Web Mercator, 109, 118

zoom scale, 138
Zoom to Full Extent, 140–41
z-values, 122